Applications of Computational Fluid Dynamics Simulation and Modeling

Edited by Suvanjan Bhattacharyya

Published in London, United Kingdom

IntechOpen

Supporting open minds since 2005

Applications of Computational Fluid Dynamics Simulation and Modeling
http://dx.doi.org/10.5772/intechopen.94665
Edited by Suvanjan Bhattacharyya

Contributors
Karthikeyan Natarajan, Chandran Suren, I. Ketut Aria Pria Utama, I. Ketut Suastika, Muhammad Luqman Hakim, Boggarapu Nageswara Rao, Vinay Atgur, Gowda Manavendra, Gururaj Pandurangarao Desai, Suvanjan Bhattacharyya, John P. Abraham, Lijing Cheng, John Gorman, Fraj Echouchene, Hafedh Belmabrouk, Marcy C. Purnell, Alejandro Rincón-Casado, Francisco José Sánchez de la Flor, Violeta Carvalho, Senhorinha Fátima Capela Fortunas Teixeira, José Carlos Teixeira, Diogo Lopes, Helder Puga, João Silva, Rui A. Lima, Vikash Kumar, Kailash Jha, Manuel Alberto Flores-Hidalgo, Diana Barraza-Jiménez, Sandra Iliana Torres-Herrera, Patricia Ponce-Peña, Carlos Omar Ríos-Orozco, Elva Marcela Coria-Quiñones, Adolfo Padilla-Mendiola, Raúl Armando Olvera-Corral, Sayda Dinorah Coria-Quiñones, Atiq Ur Rehman, Akshoy Ranjan Paul, Anuj Jain

Notice
Statements and opinions expressed in the chapters are these of the individual contributors and not necessarily those of the editors or publisher. No responsibility is accepted for the accuracy of information contained in the published chapters. The publisher assumes no responsibility for any damage or injury to persons or property arising out of the use of any materials, instructions, methods or ideas contained in the book.

First published in London, United Kingdom, 2022 by IntechOpen
IntechOpen is the global imprint of INTECHOPEN LIMITED, registered in England and Wales, registration number: 11086078, 5 Princes Gate Court, London, SW7 2QJ, United Kingdom
Printed in Croatia

British Library Cataloguing-in-Publication Data
A catalogue record for this book is available from the British Library

Additional hard and PDF copies can be obtained from orders@intechopen.com

Applications of Computational Fluid Dynamics Simulation and Modeling
Edited by Suvanjan Bhattacharyya
p. cm.
Print ISBN 978-1-83968-247-6
Online ISBN 978-1-83968-248-3
eBook (PDF) ISBN 978-1-83968-321-3

We are IntechOpen,
the world's leading publisher of
Open Access books

Built by scientists, for scientists

6,000+
Open access books available

148,000+
International authors and editors

185M+
Downloads

Our authors are among the

156
Countries delivered to

Top 1%
most cited scientists

12.2%
Contributors from top 500 universities

CLARIVATE ANALYTICS
BOOK
CITATION
INDEX
INDEXED

WEB OF SCIENCE™

Selection of our books indexed in the Book Citation Index (BKCI)
in Web of Science Core Collection™

Interested in publishing with us?
Contact book.department@intechopen.com

Numbers displayed above are based on latest data collected.
For more information visit www.intechopen.com

Meet the editor

Dr. Suvanjan Bhattacharyya is an assistant professor in the Department of Mechanical Engineering, Birla Institute of Technology and Science (BITS), Pilani, India. He completed his postdoctoral research in the Department of Mechanical and Aeronautical Engineering, University of Pretoria, South Africa, under the supervision of Prof. Josua P. Meyer. Dr. Bhattacharyya obtained his Ph.D. in Mechanical Engineering from Jadavpur University, Kolkata, India with the collaboration of Dusseldorf University of Applied Sciences, Germany. He received his master's degree from the Indian Institute of Engineering, Science and Technology, India (formerly known as Bengal Engineering and Science University). His research interest lies in computational fluid dynamics in fluid flow and heat transfer, specializing in laminar, turbulent, steady, unsteady separated flows and convective heat transfer, experimental heat transfer enhancement, solar energy, and renewable energy. He is the author and co-author of ninety papers in journals and conference proceedings. Dr. Suvanjan is listed among the world's top 2% scientists by Stanford University, USA. He has received best paper awards at several international conferences. He is an editorial board member for eleven journals and a reviewer for more than fifty others. He is currently supervising five PhD students and has several research grants to his credit from DST, DRDO, ISHRAE, and other organizations. He has also implemented several international collaborative projects in recent times. His strong collaboration with universities in South Africa led to a bilateral India–South Africa grant from the Department of Science (DST), Government of India. Dr. Bhattacharyya is currently working on a project titled "Performance Improvement of Solar Thermal Systems using Magnetic Nanofluids" in collaboration with Tshwane University of Technology, South Africa and supported by the DST.

Contents

Preface

Computational fluid dynamics (CFD) is an emerging and important tool for addressing problems in engineering, medical science, technology, and other areas. This book presents important research work related to the CFD analysis of heat exchangers, flow obstruction, flow separation, and more. It provides in-depth knowledge of the thermohydraulic and flow characteristics of various thermal and flow systems.

We have carefully selected the various research works to include in the present book. We hope our readers will love the book and gain the relevant knowledge required for their voyage of learning.

Suvanjan Bhattacharyya
Department of Mechanical Engineering,
Birla Institute of Technology and Science (BITS) Pilani,
Pilani, Rajasthan, India

Introductory Chapter: A Brief History of and Introduction to Computational Fluid Dynamics

Suvanjan Bhattacharyya, John P. Abraham, Lijing Cheng and John Gorman

1. Introduction

Werner Heisenberg: "*When I meet God, I am going to ask him two questions. Why relativity and why turbulence? I really believe he will have an answer for the first.*"

Richard Feynman: "*There is a physical problem that is common to many fields, that is very old, and that has not been solved. It is not the there is a problem of finding new fundamental particles, but something left over from a long time ago … over a hundred years. Nobody in physics has been able to analyze it satisfactorily in spite of its importance to the sister sciences. It is the analysis of circulating or turbulent fluids.*"

These quotes, both from Nobel prize winning scientists, may be apocryphal. But they nevertheless help provide context for our current understanding of fluid flow. And while these statements were reportedly made many decades ago, there is still truth to them. But at a risk of being brazen, we may now have the critical tools necessary to solve complex fluid flow problems with acceptable accuracy and fidelity. Those tools of numerical simulation are the focus of this chapter and this book.

It is not an overstatement to call numerical methods in general, and numerical simulation of fluid flow, in particular, a critical development in science in the past 100 years. In this chapter, we intend to briefly discuss the historical development of this science before quickly moving into the essential aspect of its practice. In our view, perhaps the most important aspect of numerical methods is that they provide solutions to problems that would otherwise require extensive experimentation or would otherwise be intractable. Simply put, numerical simulation opens the door for solutions to many academic and real-world problems that could otherwise not be solved.

One reason for the importance of numerical simulation for thermal-fluid problems is that the governing equations of motion are highly non-linear and coupled. That is, solutions require the simultaneous consideration of momentum (in all three coordinate directions), conservation of mass, conservation of energy (particularly for problems that involve heat transfer), and the potential for additional turbulence equations and species conservation/reaction equations. Such problems are not capable of being solved analytically. Furthermore, experimentation is often prohibitively expensive, time consuming, or impossible.

As we will see, numerical simulations of flows are important at many spatial and temporal scales. From the nanoscale to astronomical scales, from microseconds to millennia. At the large scales, climate and weather simulations are nearly ubiquitously used to make predictions. Small-scale examples include flows of fluid through microchannels or around micron-scale (or smaller objects) are representative. The breadth of scales is indicative of the wide applications this technique has been applied to.

2. A brief history of CFD

Computational Fluid Dynamics (CFD) refers to a broad set of methods that are used to solve the coupled nonlinear equations that govern fluid motion. To our best knowledge, the first attempt to calculate fluid flow was set forth by Lewis Fry Richardson, with applications for weather prediction. He envisioned a "forecast factory" that included 64,000 human "computers". Each "computer" was positioned at tiered elevations around a spherical globe, occupying computational cells that corresponded to map locations, as shown below for northern Europe. His method involved inputting weather observation data to the corresponding grid locations and then solving the forward-stepping equations.

Based on Richardson's description, the following image provides the imagined weather prediction system, commonly referred to as the "fantastic weather factory" of Lewis Richardson. Each of the red and white grid cells represents a human calculator. They are arranged across the surface of a sphere (which represents the Earth). In the center, a conductor uses spotlights to highlight calculated results at each grid cell. It was acknowledged that for such a system to work, each human calculator would be required to perform their calculations at the same speed. That is, if one human calculator was either faster or slower than its neighbors, it would send information to the neighbors at a faster or slower rate which would consequently cause numerical instability; a concept that is important even today as we will show. In the image, the blue spotlight identifies calculators that are operating too slowly, and the red spotlight identifies those that are too fast.

While the vision of Richardson is somewhat fanciful, it nevertheless provides the foundation for modern day numerical simulation. His vision was remarkably prescient.

It was not until electric computers were developed that CFD could really begin its development and the human computers of Richardson could be replaced by digital computation. This generally commenced in the 1940s with the ENIAC programmable digital computer. Small scale simulations began to appear in the scientific literature in the early 1950s (for example [1]).

3. Stability of CFD simulations and the CFD process

As indicated in the discussion of **Figures 1** and **2**, many applications of CFD are inherently unsteady. Consider for example CFD for weather forecasting where local velocity, pressure and temperatures are varying continuously. In other cases, the flow situation may initially appear steady, with steady boundary conditions. However, the flow patterns emerge as naturally unsteady. **Figure 3** provides an example of such a situation. There, steady flow approaches a small square cylinder from the left. In the downstream wake region, the flow expresses unsteady alternating vortices (the so-called Karmen vortex structure). Whenever flow experiences unsteadiness, a modeler should consider the time steps required for solution stability. Even with a situation like that of **Figure 3**, integration forward in time and stability criteria are important.

The unsteadiness, in connection with the coupled nature of the nonlinear equations, make the stability of the solutions a particular vexing issue. In fact, the issue of stability was recognized in the early 1900s and criteria for stability were soon

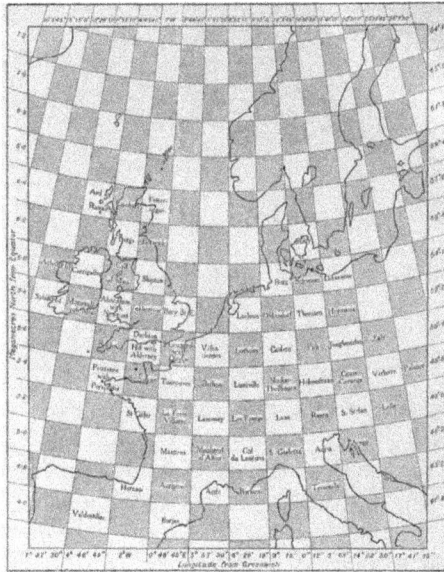

Figure 1.
Numerical grid distribution, from [2].

Figure 2.
Artist depiction of the Richardson "fantastic weather factory" Image: ink and watercolour © Stephen Conlin 1986. All Rights Reserved. Based on advice from Prof. John Byrne, Trinity College Dublin.

developed. In 1928 [3] a stability criterion was developed that provided a limitation to the time step that could be used in unsteady problems. The criterion, now termed the (Courant–Friedrichs–Lewy) CFL condition, is still used today. In short, the CFL conditions stipulates that information cannot flow entirely across a computational element in a single time step. Consequently, the local velocity of fluid, multiplied by the time step, must be smaller than the element size. This issue was highlighted in **Figure 2**, by the imagined red and blue spotlights.

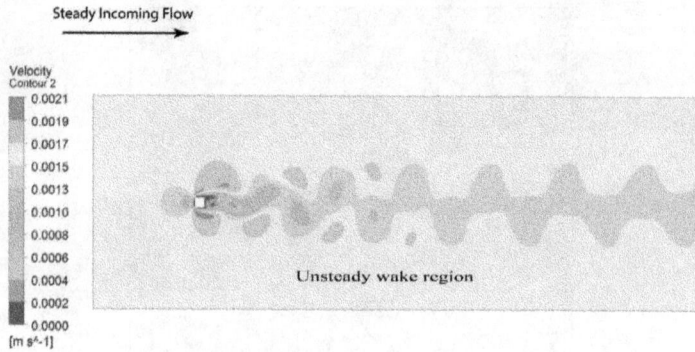

Figure 3.
An unsteady flow that results from steady boundary conditions.

The CFL criterion was developed for explicit numerical schemes but it is also used for implicit schemes as a timestep benchmark. It is worthwhile to discuss "explicit" and "implicit" numerical schemes. To aid in the discussion, **Figure 4** is provided. There, a user begins a CFD analysis by creating the flow geometry (which includes the volume occupied by the fluid). Next, the computational mesh is created (which is a collection of grid cells used to subdivide the domain.

An example of the computational mesh that was used to provide the results set forth in **Figure 3**, is shown below. These images are from the present authors' research but are typical of general CFD mesh deployments. In the figure, a series of images are provided with increasing focus on the fluid region adjacent to a square object. As seen in the series of images, the elements in the vicinity of the object are much finer than elements in further away. The use of locally refined elements is a technique to provide high accuracy in areas that are of critical importance to the analysis.

Typically, a researcher will not know *a priori* whether a particular mesh is suitable for a calculation. The requirements of the mesh depend in part on the necessary accuracy of the simulations. However, it is common for researchers to perform mesh refinement studies to ensure a level of accuracy. In a mesh refinement

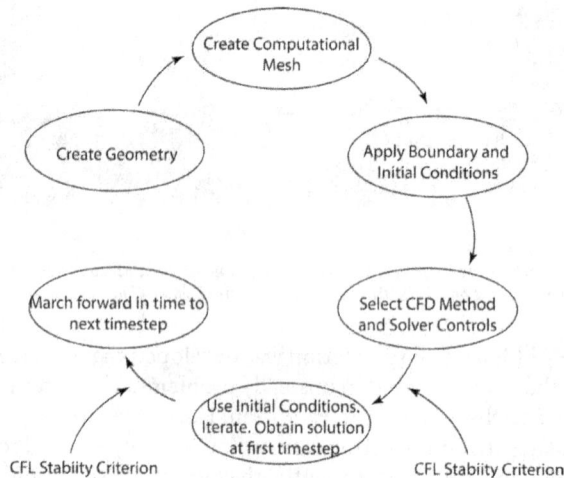

Figure 4.
The CFD process.

study, the user will refine a computational mesh and perform replicate simulations, until further refinement of the mesh does not lead to significant changes in the results. Such a situation is termed "mesh independent". It is standard practice that final solutions are mesh independent and at least within the scientific literature, mesh-independent solutions are standard. For unsteady simulations, the size of the time step is also important and similar time-step independent studies can also be performed.

With the computational mesh now created, the user moves to the next step in the process which is the application of boundary and initial conditions. Traditionally, initial conditions refer to the starting conditions of an unsteady problem. However with CFD, initial conditions are necessary even if the problem is truly steady state. For steady problems, initial conditions are the solution that commences the iterations.

Next, the numerical method and solver controls are defined. This step includes decisions such as:

- Is the flow laminar or turbulent?

- If the flow is turbulent, what turbulent model will be employed?

- Will an explicit or implicit solver be used?

- How many iterations should be performed to converge the system of equations?

- How small should the iteration-to-iteration changes be before convergence?

- Is relaxation necessary?

Next, the actual calculations can commence. The calculations involve iterating to convergence the coupled nonlinear equations. At each computational element (grid cell) the equations of mass, momentum, and energy conservation are applied. So to are the equations of turbulence. For a two-equation turbulent model, each grid cell will require seven equations (mass, three momentum, one energy, two turbulence). Consequently, a 1,000,000 element simulation will result in 7,000,000 coupled, nonlinear equations. Obviously an iteration solution strategy is required.

This iteration procedure results in a solution at the first time step. Since a time step has been taken, the CFL stability criterion is employed. Traditionally, the CFL stability criterion is enforced for explicit time-stepping schemes, but not for implicit methods. With an implicit time-stepping method, the results at the next time step are solely based on the solution at the prior time step. A result of an explicit scheme is that the equations are simpler to formulate and solve, compared to implicit methods.

As an alternative to explicit time stepping, a user may wish to use an implicit approach. With implicit algorithms, the pressure, velocity, temperature and turbulent results at a future time step depend both on the results of the prior time step as well as on the results of the future time step. Obviously, such a definition requires a more comprehensive sub-iteration procedure in order to converge to a solution, but the results are categorically stable (and therefore not subject to the CFL criterion). There are variations in implicit schemes, for example a fully implicit scheme relies only on information at the current time step. On the other hand, a Crank-Nicholson scheme relies equally upon results at a prior and future time step. Regardless of the details, implicit schemes are stable.

In our view, the stability of implicit schemes is not a strength, rather it is a weakness. The basis for this opinion is that it is possible to use a time step that is too large to achieve accurate solutions with an implicit solver, but the solution will nevertheless be stable. An unexperienced CFD research may presume that a stable solution is also an accurate solution – but this presumption is often in error. Therefore, even when using an implicit time stepping scheme, the user should pay close attention to the influence of time step size on accuracy and we recommend that the CFL criterion be applied as a guide for determining the required time step size.

It is also important to recognize that the time step size varies inversely with the element size. Consequently, when elements are made small to improve accuracy, the time steps also must become smaller to ensure convergence. Because of this, for unsteady calculations, a mesh refinement study will require more effort to solve each iteration, and more time steps are required because the time steps must be accordingly smaller.

The above discussion relates to what are often termed "numerical error". But there is another, more nuanced source of error we refer to as "modeling error". Modeling error is not related to element size or time step size, it is instead focused on the inexact input of material properties, boundary conditions, and other features of the simulation. Colloquially, we refer to "garbage in gives garbage out" and this adage is true. Insofar as inputs to the computational model deviate from a real-life situation, a user can expect differences between the simulated and actual results. In our experience, modeling errors are more significant than numerical errors. They are often much harder to diagnose and remove. Our recommendation is that CFD users pay particular attention to ensuring the inputs to their computational model match the expected inputs in real life.

4. Summary of other important CFD advances

Here we will discuss some technical innovations that have allowed CFD to become a widely available tool for researchers.

4.1 Coupling of velocity–pressure equations

Regardless of whether a simulation is steady or unsteady, an iterative process must be undertaken wherein solutions are fed back into the coupled equations and then the solutions are updated to improve accuracy. The pressure and velocity fields are coupled via the governing equations and it is challenging to calculate them simultaneously. A series of approaches was developed that began with the so-called SIMPLE algorithm (Semi-Implicit Method for Pressure Linked Equations) [4, 5]. These references provide extensive detail on the algorithm, which has been joined by a modified SIMPLER version, the SIMPLEC method, and the PISO method. All of these methods continue to be used in today's algorithms. Use of these approaches greatly improves the stability of the pressure–velocity coupling.

4.2 Multigrid calculations

Another major innovation has been the utilization of multigrid simulations to increase the speed of solution. With a multi-grid method, the governing equations are first solved on a coarse mesh. This solution is then sent to a new mesh that contains smaller elements and iterations with the new mesh are performed. The solution is again passed to a further refined mesh and the process continues until the mesh is identical with the cells that the user has defined (for example in **Figure 5**). Following a converged

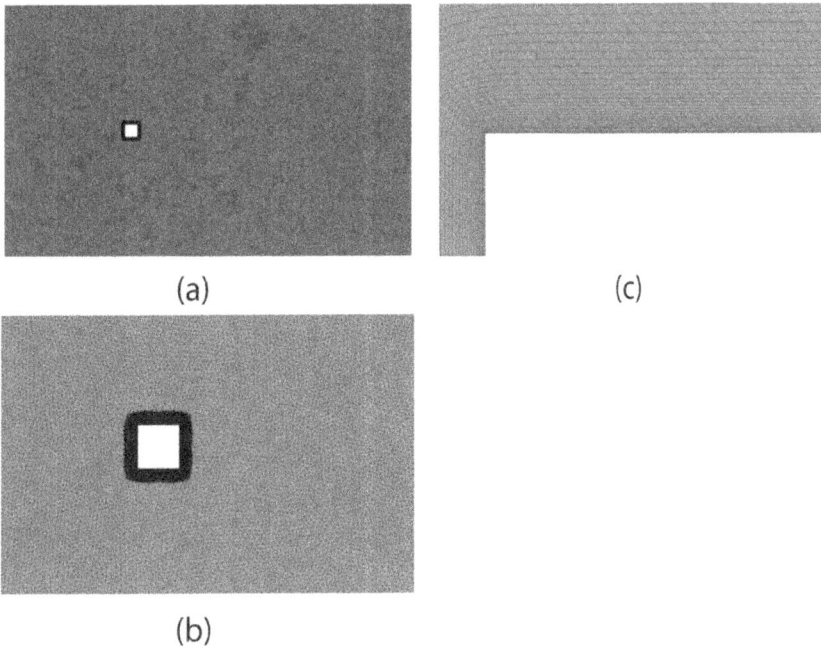

(a)   (c)

(b)

Figure 5.
A computational mesh with increased focus on the near-wall region.

solution on the smallest cells, the process is reversed, information is passed to increasingly coarse meshes until the algorithm arrives back at the initial, coarse mesh.

Through this process, solution information is transferred from the boundary conditions to the interior of the solution domain much faster than if the solution were obtained only on the finest computational mesh. It should be recognized that information is able to travel the distance of a cell size in a single iteration. If the solution domain is composed entirely of small elements, it means many iterations must be processed simply to transfer boundary information to the interior of the domain. The multi-grid approach largely solves this problem.

4.3 Parallel computing

Parallel computing refers to the simultaneous use of multiple processors during the iterative procedure. The software separates the computational mesh and parses different parts of the mesh to individual CPUs. Each CPU performs iterations on its batch of cells. The results are then collected, information is sent between the batches of cells, and the process is repeated.

To provide an example, a simulation that requires 10,000,000 elements that is solved with 10 processors would involve an allocation of approximately 1,000,000 elements to each processor. Clearly solving for 1,000,000 elements is far faster than the entire solution, and thus a time savings is achieved. On the other hand, there is a loss of efficiency in the partitioning (separating the elements into individual batches for each processor and passing information between the processors).

Parallel processing speed is measured as follows:

$$Speed\ Up\ Time = \frac{Time\ Required\ for\ Single\ Processor}{Time\ Required\ for\ Multiple\ Processors} \qquad (1)$$

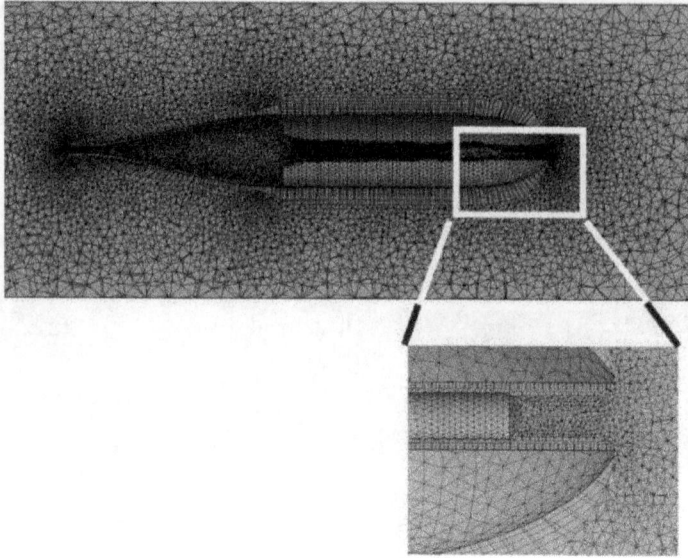

Figure 6.
Example of elements that follow curved boundaries.

Even with standard personal computers which typically have 4–8 CPU cores, a significant speed up time can be achieved. However, parallel computing is not advised for problems with a small number of elements. As a rule of thumb, each processor core should have approximately 100,000 computational elements assigned to it in order for there to be a time savings. For problems with 100,000 elements, a single processor is typically more efficient. For problems with more than approximately 200,000 elements, two processors are recommended, and so forth.

4.4 Complex element shapes

The last technical innovation that will be discussed is the advancement in element shapes. Initially, computational elements were made with simple shapes (squares and rectangles). While rectangular elements are suitable for rectangular solution domains, they are not necessarily appropriate for more complex shapes. For example, at curved boundaries, the elements would follow the boundary using a stair-step deployment. Modern day simulation algorithms generally no longer require rectangular elements. The governing equations can be solved with arbitrarily shaped elements. As a representative example, we provide a mesh that was recently used to simulate water flow past an oceanography instrument called the eXpendable BathyThermograph (XBT) As evident from the figure, a mesh has been deployed that is able to exactly follow the curved surface of the device and accurately depict the flow region (**Figure 6**).

5. Concluding remarks

We hope this brief discuss will provide valuable context for CFD users, as they read the following chapters in this book.

Author details

Suvanjan Bhattacharyya[1*], John P. Abraham[2], Lijing Cheng[3] and John Gorman[2]

1 Department of Mechanical Engineering, Birla Institute of Technology and Science Pilani, Pilani Campus, Vidya Vihar, Pilani, Rajasthan, India

2 University of St. Thomas, St. Paul, Minnesota, USA

3 Institute of Atmospheric Physics, Chinese Academy of Sciences, Beijing, China

*Address all correspondence to: suvanjan.bhattacharyya@pilani.bits-pilani.ac.in

IntechOpen

References

[1] Kawaguti W. Numerical simulation of the NS equations for flow around a circular cylinder at Reynolds number 40. Journal of the Physical Society of Japan. 1953;8:747-753.

[2] Richardson, L. Weather Prediction by Numerical Process, Cambridge University Press, 1922.

[3] Courant R, Friedrichs K, Lewy H. Uber die partiellen differenzen gleichungen der mathematischen physic. Mathematische Annalen. 1928:32-74

[4] Patankar S, Spalding D. A calculation procedure for heat, mass and momentum transfer in three-dimensional parabolic flows. International Journal of Heat and Mass Transfer. 1972;15:1787-1806.

[5] Patankar S. Numerical heat transfer and fluid flow, Taylor and Francis, 1980.

Chapter 2

Turbulence Models Commonly Used in CFD

John Gorman, Suvanjan Bhattacharyya, Lijing Cheng and John P. Abraham

Abstract

Here we provide an overview of some of the most commonly used turbulence models used in current CFD modeling. We compare the governing equations, applications of use, and results between the models. Finally, we provide our own recommendations, based on more than two decades of collaborative research.

Keywords: computational fluid dynamics, numerical simulation, turbulent flow, laminar flow, transitional flow, RANS, LES

1. Introduction

Calculation of turbulent flows is one of the most challenging problems in all of science and mathematics. Exact solutions of turbulence have bedeviled researchers for many decades and it is generally appreciated that there is no closed form solution of any fluid flow problem except the most simple laminar situations. Despite this fact, there are ways to complete calculations with sufficient accuracy so that engineering and design decisions can be made. The accuracy of turbulent calculations has gradually improved with more powerful computational resources and with improvements to numerical modeling. Here we discuss the most commonly used methods to simulate turbulent flow and discuss the strengths and weaknesses of each approach. The authors believe that particular methods are more or less appropriate for a particular situation, depending on the characteristics of the system, the computational resources available and the accuracy requirements. In this chapter, we pay particular attention to turbulence models that are most commonly used by scientists and researchers; we also provide guidance to researchers who are pondering different turbulent-modeling approaches.

2. Turbulence and CFD

The first problems handled by CFD were relatively simple, two-dimensional, incompressible, steady state situations that often were limited to laminar flows. To our best knowledge, the first three-dimensional CFD simulation was not completed until 1967 [1]. Around the same time, the very first climate models were being constructed, for modeling the circulation of fluids around the globe. Shortly thereafter, progress became much more rapid as both computational power and modeling approaches advanced. A key development was the incorporation of turbulence modeling into the CFD solutions. The first turbulence models accounted for turbulence effects through a concept termed the "eddy viscosity". Essentially, the eddy viscosity (or turbulent viscosity) reflects an apparent increase in viscosity caused by small-scale chaotic motions in a fluid. The simulations do not attempt to actually capture small scale turbulent motions, rather they approximate their effect with an increase in the fluid viscosity. As we will discuss, the concept of turbulence viscosity plays a central role in Reynolds Averaged Navier Stokes (RANS) models. As we will also show, other approaches do not rely extensively on the turbulent viscosity concept.

2.1 RANS models

The first turbulent viscosity "eddy viscosity" models were developed in the 1960s and are classified as algebraic [2, 3], one-Equation [4], or two-Equation [5–7]. The basis for two equation models was the relationship between the turbulent viscosity and local values of the turbulent kinetic energy k and turbulent dissipation, ε. Since this approach soon became the dominant method (even for today), it is worthwhile to discuss it in some detail. In essence, this group of turbulence models neglect small scale and rapid turbulent motions and use an average flow field (timewise average values in the velocities and pressure values) to estimate the effects of turbulence.

2.1.1 k-ε models

The first major effort to simulate turbulence in the context of CFD was the so-called k-ε model [5, 6]. This approach utilizes the fluctuating components of the turbulent velocity in the three coordinate directions to obtain a turbulent kinetic energy, from:

$$k = \frac{1}{2}\left(\overline{u'^2} + \overline{v'^2} + \overline{w'^2}\right) \tag{1}$$

That is, k is the additional turbulent energy that results from the time-fluctuating turbulent motions. Accompanying the turbulent kinetic energy is a turbulent dissipation ε which can be calculated as

$$\varepsilon = \frac{\kappa^{3/2}}{0.3D} \tag{2}$$

for flows in pipes with diameter D [7, 8]. The connection of turbulence kinetic energy and turbulent dissipation will be provided, following the equations of motion. In essence, the governing equations of motion are conservation of mass, which under steady conditions is:

$$\frac{\partial u_i}{\partial x_i} = 0 \tag{3}$$

conservation of momentum, written as:

$$\rho\left(u_i \frac{\partial u_j}{\partial x_i}\right) = -\frac{\partial p}{\partial x_i} + \frac{\partial}{\partial x_i}\left((\mu + \mu_t)\frac{\partial u_j}{\partial x_i}\right) j = 1, 2, 3 \tag{4}$$

and the closure equations for turbulence:

$$\rho\frac{\partial(u_i k)}{\partial x_i} = P_k + P_b - \rho\varepsilon + \frac{\partial}{\partial x_i}\left[\left(\mu + \frac{\mu_t}{\sigma_k}\right)\frac{\partial k}{\partial x_i}\right] \tag{5}$$

$$\rho\frac{\partial(u_i \varepsilon)}{\partial x_i} = \frac{\partial}{\partial x_i}\left[\left(\mu + \frac{\mu_t}{\sigma_\varepsilon}\right)\frac{\partial \varepsilon}{\partial x_i}\right] + C_1\frac{\varepsilon}{k}(P_k + C_3 P_b) - C_2\rho\frac{\varepsilon^2}{k} \tag{6}$$

The turbulent viscosity is calculated from

$$\mu_t = \rho C_\mu \frac{k^2}{\varepsilon} \tag{7}$$

The P_k is the production of turbulent kinetic energy from the shear strain rate and P_b is the production of turbulent kinetic energy from buoyancy effects. The production of turbulent kinetic energy is obtained from the time-averaged velocity field from:

$$P_\kappa = \mu_t\left(\frac{\partial u_i}{\partial x_j} + \frac{\partial u_j}{\partial x_i}\right)\frac{\partial u_i}{\partial x_j} - \frac{\partial u_\kappa}{\partial x_\kappa}\left(3\mu_t\frac{\partial u_\kappa}{\partial x_\kappa} + \rho\kappa\right) \tag{8}$$

The σ terms are corresponding Prandtl numbers for the transported variables. The values of the constants and turbulent Prandtl numbers are specific to a particular k-ε model. The k-ε approach is likely the most widely used turbulent model, even today. It is generally sufficient for flows that are wall bounded, with limited adverse pressure gradients or separation.

Traditionally, the elements are not used to capture steep velocity and temperature gradients near the wall. Rather, wall functions are employed to interpolate to the wall. Of course, the accuracy of this approach depends on the suitability of a particular wall function to a problem. For example, wall functions often fail when the flow experiences adverse pressure gradients and/or separation. On the other hand, when small elements are deployed near the wall and/or when damping equations are used to limit fluid motion in the boundary layer, integration can be performed up to the wall. In our experience, if integration is to be performed up to the wall (and wall function interpolation is avoided), the near-wall element should have a size of y+~1 for models that resolve the boundary layer. This guidance is not used for models that use the law-of-the-wall to interpolate to the wall.

A popular modification of the traditional k-ε model is the RNG (Renormalization Group) model. It was developed by [9] in an effort to handle small flow phenomenon. The mechanism of multiple scale motions is achieved by modifying the turbulent dissipation equation production term. In our experience, it has somewhat better performance than the standard k-ε particularly for rotating flows. The differences between the RNG and standard models is in the relationship between the turbulent kinetic energy, turbulent dissipation, and turbulent viscosity. With the RNG approach the turbulent viscosity is found from:

$$\mu_t = C_{\mu RNG}\rho\frac{\kappa^2}{\varepsilon} \tag{9}$$

and the new turbulent dissipation transport equation becomes:

$$\frac{\partial}{\partial x_i}(\rho u_i \varepsilon) = \frac{\partial}{\partial x_i}\left[\left(\mu + \frac{\mu_t}{\sigma_{\varepsilon RNG}}\right)\frac{\partial \varepsilon}{\partial x_i}\right] + \frac{\varepsilon}{\kappa}(C_{\varepsilon 1RNG}P_\kappa - C_{\varepsilon 2RNG}\rho\varepsilon) \tag{10}$$

With the following inputs

$$C_{e1RNG} = 1.42 - \frac{\eta\left(1 - \frac{\eta}{4.38}\right)}{\left(1 + \beta_{RNG}\eta^3\right)} \tag{11}$$

$$\eta = \sqrt{\frac{P_\kappa}{\rho C_{\mu RNG}\varepsilon}} \tag{12}$$

2.2 k-ω models

While the k-ε model has experienced success in computational modeling, it has deficiencies in some situations. In particular, the k-ε model performs suitably away from walls, in the main flow. However, it has issues in the boundary layer zone, particularly with low Reynolds numbers. Here, Reynolds numbers refer to local Reynolds numbers that decrease as one moves closer to the wall and the no-slip condition exerts its influence (rather than to the Reynolds number based on macroscopic dimensions such a pipe diameter or plate length).

A significant development in CFD was brought forward by the development of k-ω model that replaced the transport equation for ε with a specific rate of turbulence dissipation, ω [10]. The new equations are:

$$\rho\frac{\partial(u_i k)}{\partial x_i} = \rho P_k + \rho P_b - \rho\beta\omega k + \frac{\partial}{\partial x_i}\left[\left(\mu + \frac{\mu_t}{\sigma_k}\right)\frac{\partial k}{\partial x_i}\right] \tag{13}$$

$$\rho\frac{\partial(u_i\omega)}{\partial x_i} = \frac{\partial}{\partial x_i}\left[\left(\mu + \frac{\mu_t}{\sigma_\omega}\right)\frac{\partial\omega}{\partial x_i}\right] + \frac{\alpha\omega}{k}P_k - \beta\rho\omega^2 \tag{14}$$

With a turbulent viscosity calculated as:

$$\mu_t = \rho\frac{\kappa}{\omega} \tag{15}$$

2.3 Shear stress transport family of models

Recognizing that the k-ε and k-ω model each have strengths and weaknesses, a new model was proposed that uses both of these approaches in a way that harnesses their strengths [11]. This new approach, termed the Shear Stress Transport model (SST), smoothly transitions from the k-ω model near the wall to the k-ε model in the main flow. With the SST model, the governing equation for turbulent dissipation is recast into an ω form. The governing equations are:

$$\frac{\partial(\rho u_i k)}{\partial x_i} = P_k - \beta_1\rho k\omega + \frac{\partial}{\partial x_i}\left[\left(\mu + \frac{\mu_t}{\sigma_k}\right)\frac{\partial k}{\partial x_i}\right] \tag{16}$$

$$\frac{\partial(\rho u_i\omega)}{\partial x_i} = \alpha_3\frac{\omega}{\kappa}P_\kappa - \beta_2\rho\omega^2 + \frac{\partial}{\partial x_i}\left[\left(\mu + \frac{\mu_t}{\sigma_\omega}\right)\frac{\partial\omega}{\partial x_i}\right] + 2(1 - F_1)\rho\frac{1}{\sigma_{\omega 2}\omega}\frac{\partial k}{\partial x_i}\frac{\partial\omega}{\partial x_i} \tag{17}$$

and the turbulent viscosity is found from

$$\mu_t = \frac{a\rho k}{max\left(a\omega, SF_2\right)} \tag{18}$$

As before, P_k is the production of turbulent kinetic energy and ω reflects the specific rate of turbulent destruction. As noted earlier, the σ terms are turbulent Prandtl numbers associated with their subscript. The function F_1 is the aforementioned blending function that transfers the k-ω model near the wall to the k-ε model away from the wall from the wall. The S term is the magnitude of the shear strain rate.

While ostensibly, the SST model is used for fully turbulent flows, it has shown ability to capture both laminar and turbulent flow regimes [12]. However, in the next section we discuss a set of modifications to the SST models that are specifically designed to handled laminar/transitional/turbulent flow regimes that are recommended.

2.4 SST transitional models

The already discussed turbulent models were largely developed based on correlations of canonical fully turbulent flow situations (such as flows over flat plates, airfoils, Falkner-Skans flows, and flows in tubes and ducts). Of course, researchers and engineers often experience situations where the flow is partially turbulent or other situations where the flow changes so that for part of the time it is laminar and other times turbulent. Consider for example pulsatile flow wherein the fluid velocity changes sufficiently so that for parts of the flow period, different flow regimes occur. There are a number of approaches to handle these situations but with respect to the RANS models, the approaches generally utilize the concept of turbulent intermittency. Intermittency was originally defined as the percentage of time that a flow was turbulent. However, more recently, turbulent intermittency has been used as a multiplier on the rate of turbulent kinetic production [13–15].

Here we will set forth two current transitional models, both based on the SST turbulence approach. The first method involves two extra transport equations. One for the intermittency, γ, which is a multiplier to the turbulent production. The transport equation for turbulent intermittency is:

$$\frac{\partial(\rho\gamma)}{\partial t} + \frac{\partial(\rho u_i\gamma)}{\partial x_i} = P_{\gamma,1} - E_{\gamma,1} + P_{\gamma,2} - E_{\gamma,2} + \frac{\partial}{\partial x_i}\left[\left(\mu + \frac{\mu_t}{\sigma_\gamma}\right)\frac{\partial\gamma}{\partial x_i}\right] \qquad (19)$$

The P and E terms are, respectively, production and dissipation of intermittency. An additional transport equation is required for the transitional momentum thickness Reynolds number. This added equation is:

$$\frac{\partial(\rho Re_{\theta t})}{\partial t} + \frac{\partial(\rho u_i Re_{\theta t})}{\partial x_i} = P_{\theta t} + \frac{\partial}{\partial x_i}\left[\sigma_{\theta t}(\mu + \mu_t)\frac{\partial Re_{\theta t}}{\partial x_i}\right] \qquad (20)$$

Together, solution to Eqs. (19) and (20) determine the local state of turbulence. They result in an intermittency that takes values between 0 and 1. For fully laminar flow, $\gamma = 0$ and the model reverts to a laminar solver. When $\gamma = 1$, the flow is fully turbulent. The turbulent production then is then multiplied by the local value of the intermittency, γ. Interested readers are invited to review the development of this model, including implementation for problems that involve heat transfer [16–22].

Recently, the above two-equation model was modified to reduce the two transitional transport equations to a single Equation [23] and that approach was later adapted by [24] to accurately solve for situations in confined pipe/duct/tube flows. Essentially, Eqs. (19) and (20) are replaced by a single intermittency equation which is:

$$\frac{\partial(\rho u_i\gamma)}{\partial x_i} = P_\gamma - E_\gamma + \frac{\partial}{\partial x_i}\left[\left(\mu + \frac{\mu_t}{\sigma_\gamma}\right)\frac{\partial\gamma}{\partial x_i}\right] \qquad (21)$$

As with the two-equation approach, the intermittency factor γ will take on values between 0 and 1. Also, as before, The P and E terms represent, respectively, the production and destruction in local value of intermittency.

For these intermittency models, the onset of turbulence is calculated by a series of correlation functions. In particular, a local value of the critical Reynolds number is determined from

$$Re_{\theta c} = C_{TU1} + C_{TU2}\, exp\left[-C_{TU3}Tu_L F_{PG}(\lambda_{\theta L})\right] \tag{22}$$

Eq. (22) is used to identify the location of laminar-turbulent transition. It is based on the local value of the momentum layer thickness. The C terms are correlation constants and are based on comparison of numerically simulated results with experimentation. An important term in Eq. (22) is the local value of the mid-boundary-layer turbulence intensity (Tu_L). This value is attained at the midpoint of the boundary layer as an output from an empirical formulation based on experimentation.

Local production of intermittency is calculated from:

$$P_\gamma = F_{length} \cdot \rho \cdot S \cdot \gamma \cdot F_{onset} \cdot [1 - \gamma] \tag{23}$$

As we have already noted, the term S is the shear strain rate. A new term that appears in Eq. (23) is the so-called onset transition term (F_{onset}) which is calculated using the following set of equations.

$$F_{onset} = MAX\left[min\left(\frac{Re_V}{2.2 \cdot Re_{\theta c}}, 2\right) - max\left(1 - \left(\frac{R_T}{3.5}\right)^3, 0\right), 0\right] \tag{24}$$

$$Re_V = \rho d_w^2 S/\mu \tag{25}$$

$$Re_T = \rho \kappa / \mu w \tag{26}$$

Similarly, the local rate of destruction of intermittency is found by:

$$E_\gamma = 0.06 \cdot \rho \cdot \Omega \cdot \gamma \cdot F_{turb} \cdot [50\gamma - 1] \tag{27}$$

$$F_{turb} = e^{-(R_T/2)^4} \tag{28}$$

$$P_\kappa = \mu_t \cdot S \cdot \Omega \tag{29}$$

We have already noted that these transitional turbulence models were initially developed for external boundary layer flows (flat plate boundary layers, airfoil flows, Falkner-Skans flows, etc.). Insofar as we have adopted them for internal flow, some modification was required. We recommend, at least for flows through pipes, tubes, and ducts, that the initial constants determined in [23] be replaced by alternative values from [24].

While we recommend the above approach for solving transitional flow problems, this area of research is also heavily studied by other researchers who have provided alternative approaches to handle such flows. We cite them here for readers who are interested in those alternative but complementary viewpoints [25–33].

2.5 Reynolds-stress models

Reynolds stress models (RSM) are quite different from the RANS approach that was just discussed. For RSMs, transport equations are used for all components of the Reynolds stress tensor and an eddy viscosity is not utilized. These models are expected to be superior for situations with non-isotropic turbulence and flows with significant components of transport in three directions. There are a number of RSM versions, some of which will be discussed here. The so-called SSG-RSM model employed here utilizes the following momentum transport equation:

$$\rho \frac{\partial u_j}{\partial t} + \rho \frac{\partial}{\partial x_i} (u_i u_j) = -\frac{\partial p'}{\partial x_j} + \mu \frac{\partial}{\partial x_i} \left(\frac{\partial u_j}{\partial x_i} + \frac{\partial u_i}{\partial x_i} \right) - \rho \frac{\partial}{\partial x_i} (\overline{u_i u_j}) + B_j \qquad (30)$$

The second-to-last term on the right-hand side represents the Reynolds stresses. There is a pseudo-pressure term p' that is calculated from the local static pressure p and local velocity gradient from the following expression.

$$p' = p + \frac{2}{3} \mu \frac{\partial u_k}{\partial x_k} \qquad (31)$$

The Reynolds stresses are calculated by a collection of six equations for all directional possibilities. The transport equations for Reynolds stresses are:

$$\rho \frac{\partial \overline{u_j u_i}}{\partial t} + \rho \frac{\partial}{\partial x_k} (u_k \overline{u_i u_j}) - \frac{\partial}{\partial x_k} \left[\left(\mu + \frac{2}{3} \rho C_s \frac{k^2}{\varepsilon} \right) \frac{\partial \overline{u_i u_j}}{\partial x_k} \right] = P_{ij} - \frac{2}{3} \delta_{ij} \rho \varepsilon + \Phi_{ij} + P_{ij,b} \qquad (32)$$

We note that a turbulence dissipation term, ε, appears in Eq. (32) and it has to be solved from its own transport equation. We refer readers to [34, 35] for more details.

A modification to the above is realized from the Baseline RSM (BSL RSM) model. It differs from the SSG RSM in that the transport equation for ε is replaced by a transport equation for ω. The new equation is:

$$\rho \frac{\partial \omega}{\partial t} + \rho \frac{\partial (u_k \omega)}{\partial x_k} = \frac{\partial}{\partial x_k} \left[\left(\mu + \frac{\mu_t}{\sigma_{\omega 3}} \right) \frac{\partial \omega}{\partial x_k} \right] + \alpha_3 \frac{\omega}{k} P_k - \beta_3 \rho \omega^2 + (1 - F_1) \frac{2\rho}{\sigma_{\omega 2}} \frac{1}{\omega} \frac{\partial k}{\partial x_k} \frac{\partial \omega}{\partial x_k}$$
$$+ P_{\omega b} \qquad (33)$$

This approach blends between two different models that are used near the wall and alternatively away from the wall. The modeling is accomplished using a weighting function, similar to the SST:

$$\phi_3 = F_1 \cdot \phi_1 + (1 - F_1) \cdot \phi_2 \qquad (34)$$

Where the symbols ϕ correspond to any particular transport variable in the near wall and far wall regions. Various constants change their values in the two regions, so that:

The constants near the wall:

$$\sigma_k = \sigma_\omega = 2, \beta = 0.075, \alpha = 0.553 \qquad (35)$$

The constants away from the wall:

$$\sigma_k = 1, \sigma_\omega = 1.168, \beta = 0.0828, \alpha = 0.44 \qquad (36)$$

The last RSM version to be discussed is the Explicit Algebraic RSM (EARSM). This approach includes a non-linear relationship between the local values of the Reynolds stresses and the vorticity tensors. It is focused on flows with secondary motions and curvature [36]. The local values of the Reynolds stresses are calculated using an anisotropy tensor which is based on algebraic equations [36]. This is contrasted with RSM approaches that solve for the Reynolds stress components using differential transport equations. The approach is to use higher order terms for many of the flow phenomena. It was designed to handle secondary flow situations and flows with extensive curvature and rotation. The governing equations are complex and lengthy and for brevity sake, we refer interested readers to [36].

2.6 Scale adaptive models

So far, we have presented RANS-based models that perform conservation calculations at each grid element. If turbulence is present, the impact of turbulence appears via the eddy viscosity. Traditionally users either *a priori* specify that the flow is laminar (so no eddy viscosity included) or the flow is turbulent (in which case an eddy viscosity is determined and applied throughout the flow field). The recent development of transitional modeling frees the researcher from having to *a priori* predict the level of turbulence. With transitional modeling, the numerical code automatically reverts to laminar flow in areas with low Reynolds numbers and also automatically becomes a turbulent model in areas where the Reynolds number is larger.

Regardless of the method that is selected, the coupled equations are solved for each computational element and the turbulent viscosity is applied to the fluid in the element under consideration.

In contrast to this approach, there is another major group of computational techniques that are termed "scale adaptive models". These are models that resolve part of the turbulent motions but model flow features that are smaller than the element size. Since there is less modeling and more actual resolution of fluid motion, one might expect the scale-adaptive models to be more accurate than RANS; and there are cases where that is so (particularly for free shear flows, swirling flows, boundary layer separation, and jets). However, the RANS approach can be more accurate than scale-adaptive methods in some situations, including wall bounded flows. Also, RANS is less computationally expensive because the eddy viscosity provides the link to the time-averaged flow field and the local turbulence with a very simple calculation. In fact, for even problems of modest complexity, scale adaptive models are more time consuming.

There are a number of established and new Scale-Adaptive Models that are used in CFD simulations. We will not be exhaustive in this section by covering all the existing models, rather we will focus on some of the models we think are most useful and representative. Interested readers are directed to an excellent comprehensive discussion provided by [34, 37].

2.6.1 Scale-adaptive SST models

One of the primary decisions that models are faced with is whether to perform calculations in steady or unsteady mode. Typically with numerical simulation, unsteadiness is driven by either timewise changes in boundary conditions or it is related to unsteady phenomena that occur in an otherwise steady scenario. A classic example is the Karmen Vortex Street that occurs in a wake region of a blunt object. **Figure 1**, shown below, illustrates this phenomenon.

Researchers have often conjectured that if a RANS model is performed with sufficiently small elements and time steps, the unsteady features of the flow would naturally be resolved. But in fact, this is not true. It is important to note that steady state calculations using RANS models will often provide very accurate information

Steady Incoming Flow

Velocity
Contour 2

0.0021
0.0019
0.0017
0.0015
0.0013
0.0010
0.0008
0.0006
0.0004
0.0002
0.0000
[m s^-1]

Unsteady wake region

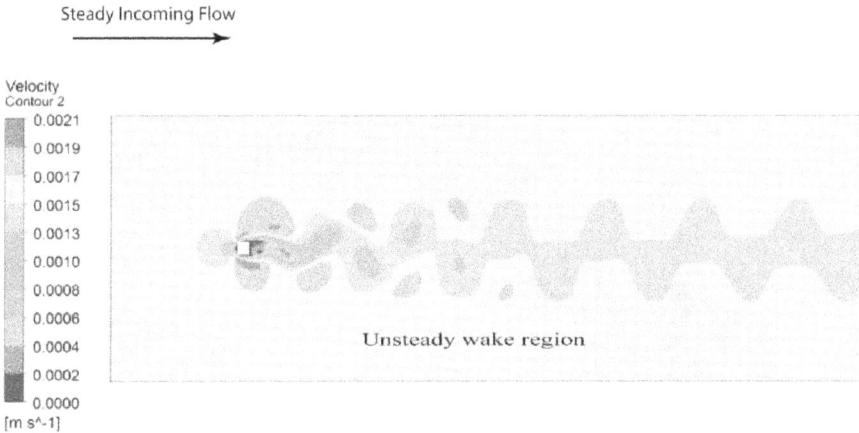

Figure 1.
Unsteady wake region, even though oncoming flow is steady state.

about averaged quantities (like drag), these simulations will miss details in the rapidly fluctuating downstream wake region. This issue was explored in depth in [35] where time-averaged results of drag obtained from unsteady RANS simulations were compared with calculations from steady RANS calculations (using the SST transitional model that was previously described). It was found that the steady state calculations were able to accurately capture drag forces but were only partially adept at capturing vortex movement in the downstream wake region.

With this discussion as background, it is now time to turn attention to the governing equations of scale-adaptive RANS models. The model to be discussed here uses the SST approach for the underlying governing equations (in the literature it is often termed the SAS-SST model). The scale-adaptive approach modifies the ω transport equation based on [37]. In particular, a new transport equation is presented that incorporates the turbulent length scale L and is set forth here:

$$\Phi = \sqrt{k}L \qquad (37)$$

and

$$\frac{\partial \Phi}{\partial t} + \rho \frac{\partial U_i \Phi}{\partial x_i} = \frac{\Phi P_k}{k} \left(C_1 - C_2 \left(\frac{L_t}{L_k} \right)^2 \right) - \rho C_3 k + \frac{\partial}{\partial x_i} \left(\frac{\mu_t}{\sigma_\Phi} \frac{\partial \Phi}{\partial x_i} \right) \qquad (38)$$

Values of the various constants can be found in [34, 37] and are not repeated here for brevity. The term L_t is a novel modification; it refers to the von Karmen length scale. **Figures 2** and **3** are provided that show a comparison of downstream wake regions for an unsteady RANS calculation using the SST model (**Figure 2**) and a simulation using the scale-adaptive SST modification. Results are obtained from [34]. It can be seen that the standard SST model does capture a periodic release of eddies from the downstream side of a circular cylinder (shown in blue). In both images, the flow is left-to-right. The color legend is keyed to the local values of the turbulent length scale. Clearly the scale-adaptive approach provides a much wider range of turbulent eddy sizes.

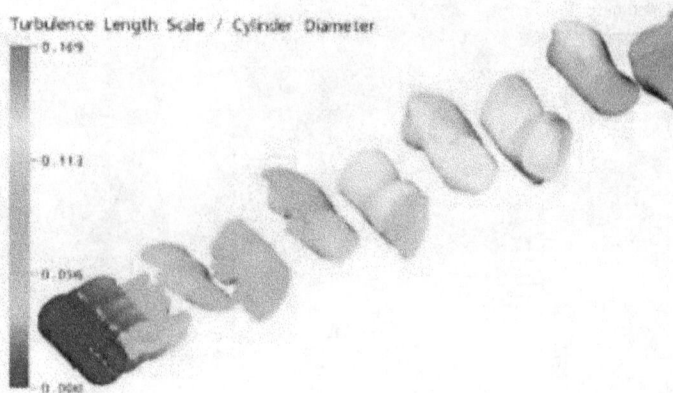

Figure 2.
Calculations of turbulent length scale for flow over a circular cylinder, based on an unsteady SST model.

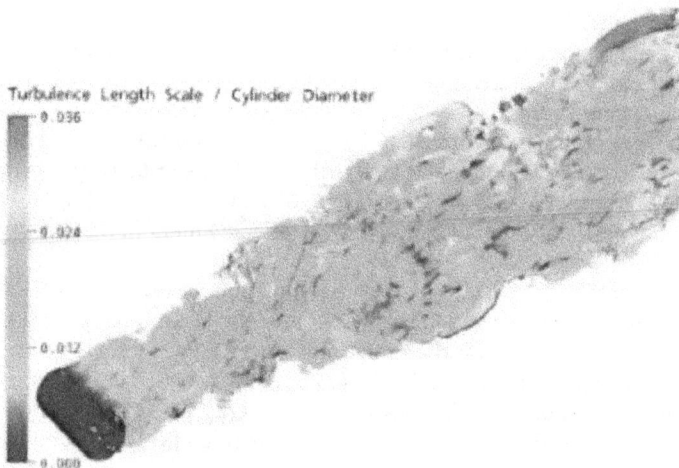

Figure 3.
Calculations of turbulent length scale for flow over a circular cylinder, based on a scale-adaptive unsteady SST model.

2.6.2 LES WALE model

Another common approach to dealing with these types of problems is based on the so-called "large eddy simulation". To the best knowledge of the authors, the first articulation of a LES model was [38] and the models have been updated in the intervening decades. Here we focus on one popular and current LES method (the Wall-Adaptive Local Eddy, or WALE LES model). The general processes of LES modeling are the same, regardless of which variant is used. LES models involve the filtering of eddies that are smaller than the size of the computational elements. The algorithm incorporates an eddy viscosity for flow scales that are not resolved.

For this model, the tensor-form of the Navier Stokes equations is:

$$\frac{\partial}{\partial t}\left(\overline{\rho u_i}\right) + \left(\frac{\partial \overline{\rho u_i u_j}}{\partial x_i}\right) = -\frac{\partial \overline{p}}{\partial x_j} + \frac{\partial}{\partial x_i}\left(\mu\left(\frac{\partial \overline{u_i}}{\partial x_j} + \frac{\partial \overline{u_j}}{\partial x_i}\right)\right) + \frac{\partial \tau_{ij}}{\partial x_i} \tag{39}$$

where τ_{ij} is the small-scale stress defined as

$$\tau_{ij} = -\overline{\rho u_i u_j} + \overline{\rho u_i} \overline{u_j} = 2\mu_{sgs} \overline{S}_{ij} + \frac{1}{3}\delta_{ij}\tau_{kk} \tag{40}$$

And the $\overline{S}_{ij}$ term indicates the strain rate tensor for large scale motions. The small-scale eddy viscosity μ_{sgs} is found from

$$\mu_{sgs} = \rho(C_w\Delta)^2 \frac{\left(S_{ij}^d S_{ij}^d\right)^{3/2}}{\left(\overline{S}_{ij}\overline{S}_{ij}\right)^{5/2} + \left(S_{ij}^d S_{ij}^d\right)^{5/4}} \tag{41}$$

The term C_w is a constant and the symbol Δ = (element volume)$^{1/3}$. The tensor S_{ij}^d is calculated from the strain-rate and vorticity tensors, as shown here

$$S_{ij}^d = \overline{S}_{ik}\overline{S}_{kj} + \overline{\Omega}_{ik}\overline{\Omega}_{kj} - \frac{1}{3}\delta_{ij}\left(\overline{S}_{mn}\overline{S}_{mn} - \overline{\Omega}_{mn}\overline{\Omega}_{mn}\right) \tag{42}$$

And the vorticity tensor $\overline{\Omega}_{ij}$ is defined as

$$\overline{\Omega}_{ij} = \frac{1}{2}\left(\frac{\partial \overline{u}_i}{\partial x_j} + \frac{\partial \overline{u}_j}{\partial x_i}\right) \tag{43}$$

3. Results from various CFD model calculations

Now that the main CFD models have been presented, we turn attention to comparisons of the results from different models. There are comparisons available in [7, 8, 34, 35, 37, 39–46] and a very small subset of those comparisons will be provided here. We have selected the classic problem of flow over a square blockage. This canonical problem has the features that elucidate the strengths and weaknesses of the particular models. For instance, some important parameters relate to the time-averaged interactions between the fluid and the solid structure (drag force). Also, there are significant unsteady phenomena, particularly in the wake region that provide a challenging test for the models. In addition, this is a problem with extensive experimental work that will serve as the basis for evaluating the results. To begin we refer to **Figure 4** which shows the solution domain (similar to [35]).

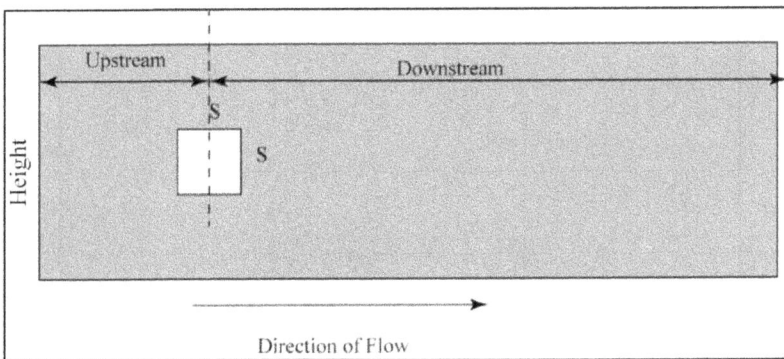

Figure 4.
Geometry for flow over a square cylinder.

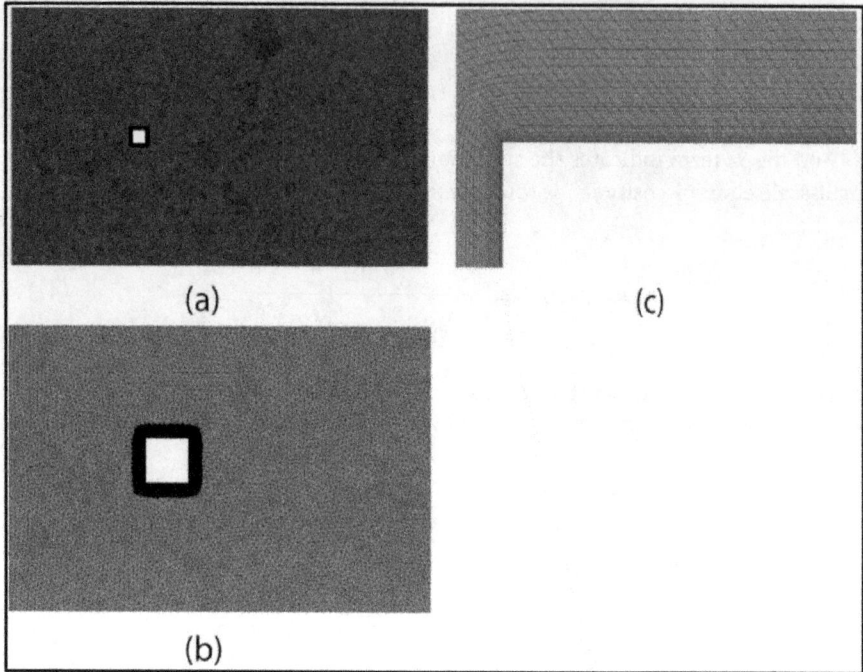

Figure 5.
Computational mesh used for square cylinder simulation.

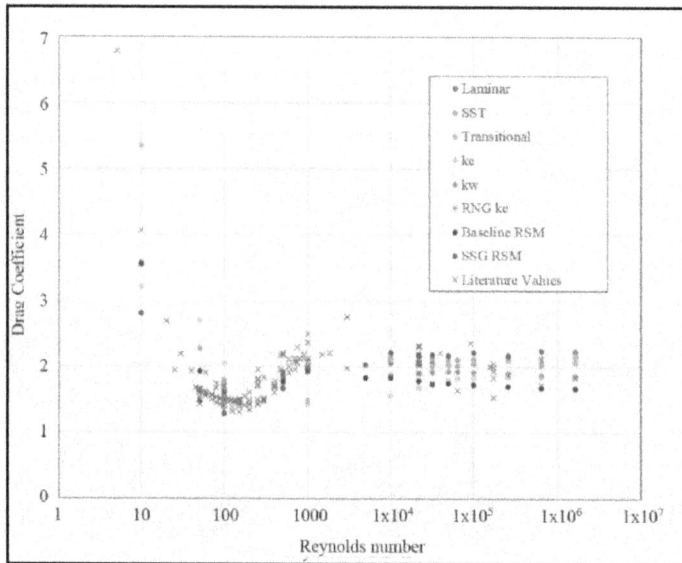

Figure 6.
Drag coefficients for flow over a square cylinder and comparison with experiments. Reynolds numbers range from 1 to 10,000,000.

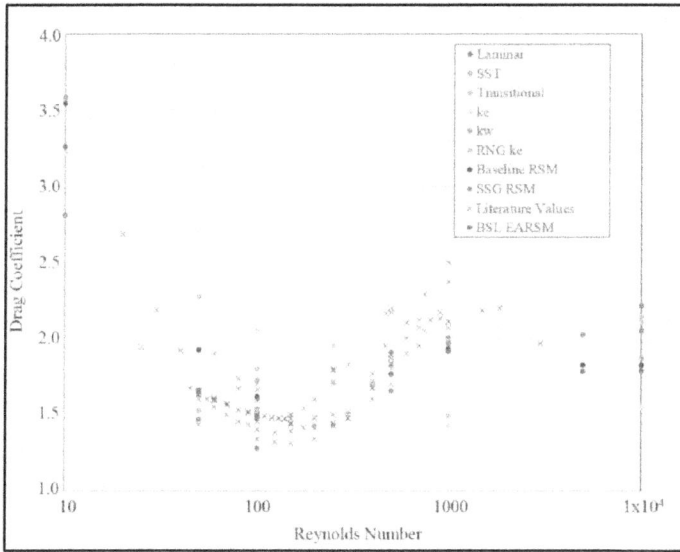

Figure 7.
Drag coefficients for flow over a square cylinder and comparison with experiments, Reynolds numbers ranging between 10 and 10,000.

A number of computational meshes were used and an example mesh is shown in **Figure 5**. The images are provided in a series of increasing magnification. Image (a) is the most global view, part (b) is focused on the square obstruction, and image (c) reveals details of the elements in the near-wall region, near a corner of the cylinder.

With this mesh, we present results for a large number of computational methods. We note here that in reality appropriate meshes may differ depending on the turbulence model that is used. For instance, a mesh that is suitable for a k-w simulation may not be appropriate for SST, and vice versa. We recommend that mesh independent studies be carried out for each turbulence mode that is employed. The results, set forth in **Figures 6** and 7, provide the drag coefficient on the square cylinder (large aspect ratio). Each model has its own color. Literature-based values from experiments are also included (shown as gray x symbols).

In the above calculations, which were first set forth in [35], the SST and transitional-SST models were most accurate (when compared with existing experiments) for calculating the drag coefficient. On the other hand, since these approaches were RANS, they lose some local detail and flow structure. For example, in **Figure 8** which is provided below, we show velocity vectors, overlaid atop a velocity contour image. It is evident from the upper part of the figure that there are the expected stagnation locations at the leading edge, and in the wake region. There is also a slow-moving recirculation zone above and below the cylinder that are a result of flow separation at the leading corners. However, the lower images show a focus on the flow patterns at the leading edge. It is seen that with the SST RANS model, there are no small-scale eddies at this location. But for the LES model, there are two LES results that are obtained at two different instances in time. These sequential images show the time-varying flow field. While a RANS model like the SST is excellent for full-body drag, it does not capture some small flow structures. Researchers thus need to consider their computational needs before selecting a CFD model.

Velocity Results - SST model

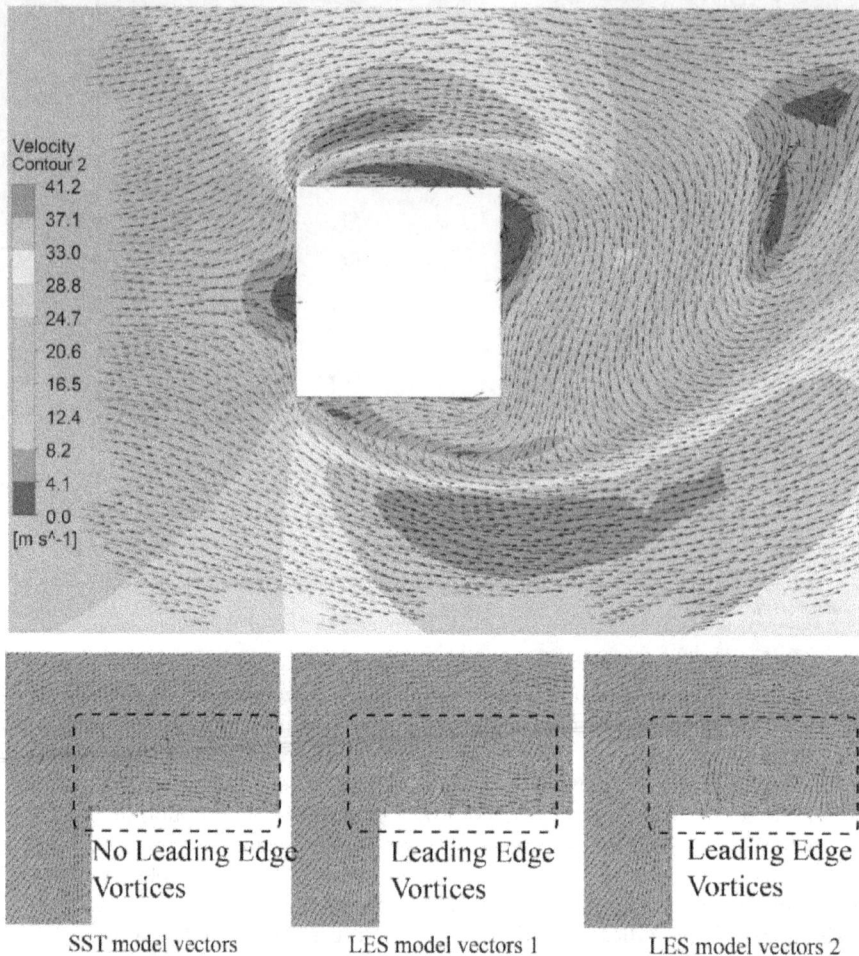

Figure 8.
(upper image) velocity contour and vectors for SST model and (lower image) side-by-side comparison of leading-edge flow for SST and WALE LES models.

The last result to be presented is shown in **Figure 9**. There, instantaneous results are displayed for the SST model. There, clearly, the unsteady nature of flow in the downstream wake region are evident. If the simulation of **Figure 9** was carried out with a steady state SST solver, there would still be timewise changes in the flow field but they would have a different frequency than the unsteady calculations.

In order to elucidate the iteration-by-iteration fluctuations in drag that result from a steady state solver (compared to an unsteady simulation), **Figure 10** is prepared. This figure shows the timewise (iteration wise) fluctuations in drag force on the square cylinder first with a steady state SST solution and then with a truly unsteady solution. The steady state results are calculated using a "false transient" approach wherein the algorithm steps forward to new iterations using a non-physical time. The figure has two call outs that provide focus on different parts of the graph. The important conclusion is that the average value of unsteady fluctuations of drag obtained by the steady state algorithm are an excellent match that that

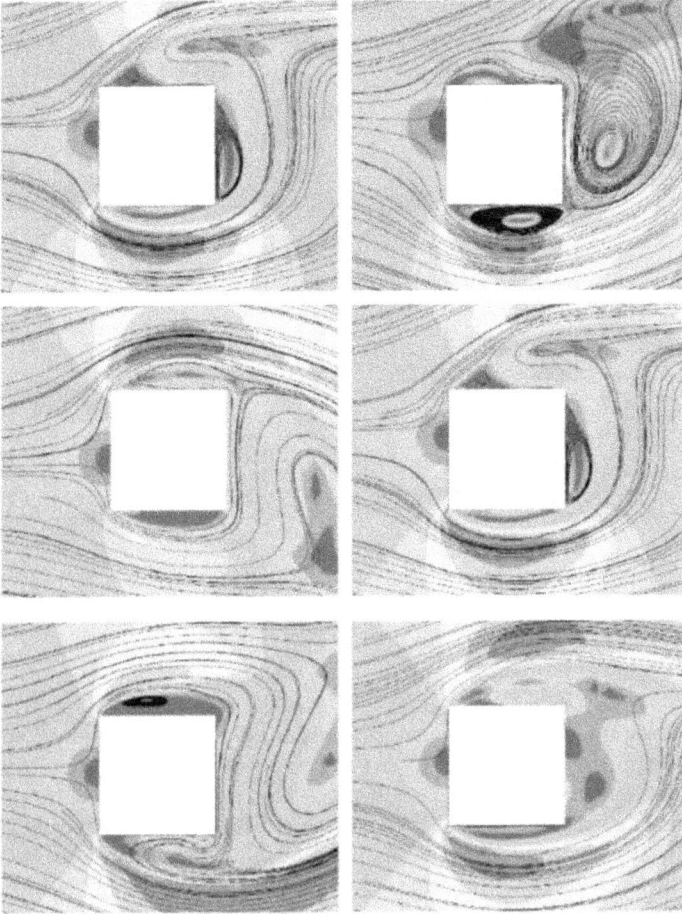

Figure 9.
Streamline patterns and velocity contours for Re = 100,000 flow over a square cylinder. Images at a sequence of time instances, using SST model.

attained from the unsteady calculations. On the other hand, the period is very different between the two.

4. Concluding remarks

This chapter has presented a brief overview of a large number of turbulence models. While there is no "correct" turbulence model, there are models that are better suited for particular situations.

For flows that are truly laminar with no regions of intermittency or turbulence, a laminar solver can be used. However, if there is a potential for any turbulent flow, caution is warranted. For flows that are fully turbulent, particularly wall bounded flows, the SST model is recommended. In our experience it is more able to capture flow phenomena compared to other RANS models. It also has excellent performance for a wide range of thermal-transport situations.

If regions of mixed flows (laminar/transitional/turbulent) are expected, of if the flows might change in time (pulsatile flows for example), the SST transitional

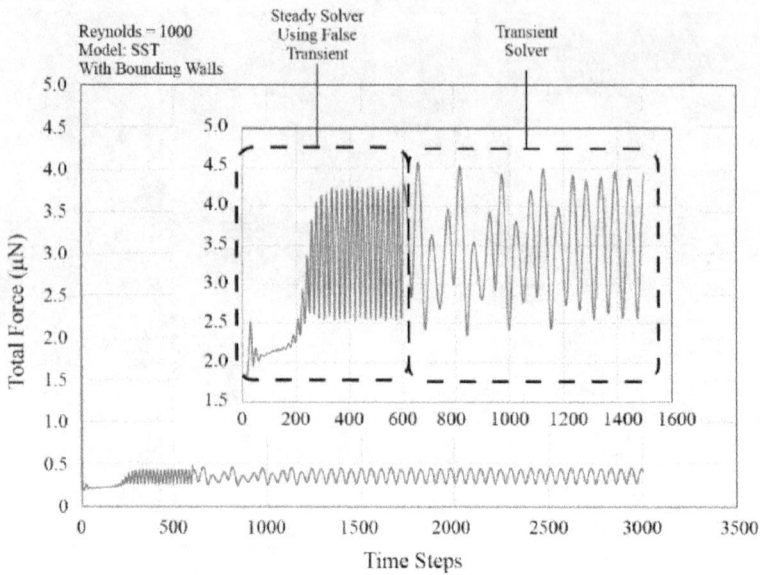

Figure 10.
SST solution that began as steady state and then was changed to unsteady.

model is recommended. This new approach is rapidly becoming more common in the CFD community and could replace fully turbulent models in the future.

For situations where small scale and short-lived flow must be captured, we recommend the scale-adaptive SST model or the LES model. They are more computationally expensive but the scale adaption enables small features to be calculated. We also direct readers to two further excellent resources [47, 48] for more in depth discussion.

Author details

John Gorman[1], Suvanjan Bhattacharyya[2], Lijing Cheng[3] and John P. Abraham[1*]

1 University of St. Thomas, St. Paul, Minnesota, USA

2 Department of Mechanical Engineering, Birla Institute of Technology and Science Pilani, Pilani, Rajasthan, India

3 Institute of Atmospheric Physics, Chinese Academy of Sciences, Beijing, China

*Address all correspondence to: jpabraham@stthomas.edu

IntechOpen

References

[1] Hess J, Smith A. Calculation of potential flow about arbitrary bodies. Progress in Aerospace Sciences. 1967;8: 1-138.

[2] Smith A, Cebeci T. Numerical simulation of the turbulent boundary layer equations, Douglass Aircraft Division Report DA 33735, 1967.

[3] Baldwin B, Lomax H. Thin layer approximation and algebraic model for separated turbulent flows. AIAA paper no. 78-257. 1978.

[4] Spalart P, Allmaras S. A one-equation turbulence model for aerodynamic flows. AIAA paper no. 92-0439, 1992.

[5] Jones W, Launder B. The prediction of laminarization with a two equation model of turbulence. International Journal of Heat and Mass Transfer. 1972; 15:301-314

[6] Launder B, Spalding D. The numerical computation of turbulent flows. Computer Methods in Applied Mechanics and Engineering. 1974;3: 269-289.

[7] Gorman J, Sparrow E, Abraham J, Minkowycz W. Evaluation of the efficacy of turbulence models for swirling flows and the effect of turbulence intensity on heat transfer. Numerical Heat Transfer Part B: Fundamentals. 2016;70:485-602

[8] Sparrow E, Gorman J, Abraham J, Minkowycz W. Validation of turbulence models for numerical simulation of fluid flow and convective heat transfer. Advances in Heat Transfer. 2017;49: 397-421.

[9] Yakhot V, Orszag S, Thangam S, Gatski T, Speziale C. Development of turbulence models for shear flows by a double expansion. Physics of Fluids. 4; 1510;1992.

[10] Wilcox D. Multiscale model for turbulence flows. AIAA Journal. 1988; 26:1311-1320.

[11] Menter F. Two-equation eddy-viscosity turbulence models for engineering applications. AIAA Journal. D1994;32:1598-1605.

[12] Sparrow E, Tong J, Abraham J. Fluid flow in a system with separate laminar and turbulent zones. Numerical Heat Transfer A. 2008;53:341-353.

[13] Menter F, Esch T, Kubacki S. Transition Modelling Based on Local Variables. In: 5th Int. Symposium on Engineering Turbulence Modeling and Measurements; 2002: Mallorca, Spain.

[14] Menter F, Langtry R, Likki S, Suzen Y, Huang P, Volker S. A Correlation-Based Transition Model Using Local Variables, Part i – Model Formulation. In: Proceedings of ASME Turbo Expo Power for Land, Sea, and Air; 14–27 June 2004: Vienna, Austria.

[15] Menter F, Langtry R, Likki S, Suzen Y, Huang P, Volker S. A. In: Proceedings of ASME Turbo Expo Power for Land, Sea, and Air. 14–17 June 2004: Vienna, Austria.

[16] Abraham J, Tong J, Sparrow E. Breakdown of laminar pipe flow into transitional intermittency and subsequent attainment of fully developed intermittent or turbulent flow. Numerical Heat Transfer B. 2008; 54:103-115.

[17] Abraham J, Sparrow E, Tong J, Heat transfer in all pipe flow regimes – Laminar, transitional/intermittent, and turbulent. International Journal of Heat and Mass Transfer. 2009;52:557-563.

[18] Minkowycz W, Abraham J, Sparrow E. Numerical simulation of

laminar breakdown and subsequent intermittent and turbulent flow in parallel plate channels: Effects of inlet velocity profile and turbulence intensity. International Journal of Heat and Mass Transfer. 2009;523: 4040-4046.

[19] Lovik R, Abraham J, Minkowycz W, Sparrow E. Laminarization and turbulentization in a pulsatile pipe flow. Numerical Heat Transfer A. 2009;56: 861-879.

[20] Abraham J, Sparrow E, Tong J, Bettenhausen D. Internal flows which transist from turbulent through intermittent to laminar. International Journal of Thermal Sciences. 2010;49: 256-263.

[21] Abraham J, Sparrow E, Minkowycz W. Internal-flow Nusselt numbers for the low-Reynolds number end of the laminar-to-turbulent transition regime. International Journal of Heat and Mass Transfer. 2011;54: 584-588.

[22] Gebreegziabher T, Sparrow E, Abraham J, Ayorinde E, Singh T. High-frequency pulsatile flows encompassing all flow regimes. Numerical Heat Transfer A. 2011;60:811-826.

[23] Menter F, Smirnov P, Liu T, Avancha R. A one-equation local correlation based transition model. Flow, Turbulence and Combustion. 2015;95:583-619.

[24] Abraham J, Sparrow E, Gorman J, Zhao Y, Minkowycz W. Application of an intermittency model for laminar, transitional, and turbulent internal flows. Journal of Fluids Engineering. 2019;141: paper no. 071204.

[25] Ghajar A, Madon K. Pressure drop measurements in the transition region for a circular tube with three different inlet configurations. Exp Therm. Fluid Sci. 1992;5:129-135.

[26] Tam L, Ghajar A. Effect of inlet geometry and heating on the fully developed friction factor in the transition region of a horizontal tube. Exp Therm. Fluid Sci. 1997;15:52-64.

[27] Tam H, Tam H, Ghajar A, Ng, W, Wong I, Leong K, Wu C. The effect of inner surface roughness and heating on friction factor in horizontal micro-tubes. Proceedings of ASME-JSME-KSME joint fluids Engineering. 2011; 24-29.

[28] Tam H, Tam L, Ghajar A. Effect of inlet geometries and heating on the entrance and fully-developed friction factors in the laminar and transition regions of a horizontal tube. Exp. Therm. Fluid Sci. 2013;44:680-696.

[29] Tam H, Tam L, Ghajar A, Sun C, Leung H. Experimental investigation of the single-phase friction factor and heat transfer inside the horizontal internally micro-fin tubes in the transition region. Proceedings of ASME-JSME-KSME joint fluids Engineering. 2011;24-29.

[30] Tam H, Tam L, Ghajar A, Sun C, Lai W. Experimental investigation of single-phase heat transfer in a horizontal internally micro-fin tube with three different inlet configurations. Proceedings of the ASME 2012 Summer Heat Transfer, 2012.

[31] Tam H, Tam L, Ghajar A, Sun C, Leung H. 2011. Experimental investigation of the single-phase friction factor and heat transfer inside the horizontal internally micro-fin tubes in the transition region. Proceedings of ASME-JSME-KSME Joint Fluids Engineering 2011;24-29.

[32] Everts M, Meyer J. Relationship between pressure drop and heat transfer of developing and fully developed flow in smooth horizontal circular tubes in the laminar, transitional, quasi-turbulent and turbulent flow regimes.

International Journal of Heat and Mass Transfer. 2018;117:1231-1250.

[33] Everts M, Meyer J. Heat transfer of developing and fully developed flow in smooth horizontal tubes in the transitional flow regime. International Journal of Heat and Mass Transfer. 2018;117:1331-1351.

[34] Menter F. Best Practices: Scale-Resolving Simulations in ANSYS CFD, Published by ANSYS, 2015.

[35] Olsen L, Abraham J, Cheng L, Gorman J, Sparrow E, Summary of forced-convection fluid flow and heat transfer for square cylinders of different aspect ratios ranging from the cube to a two-dimensional cylinder. Advances in Heat Transfer. 2019;51:351-457.

[36] Wallin S, Johansson A. An explicit algebraic Reynolds stress model for incompressible and compressible turbulent flows. Journal of Fluid Mechanics. 2000;403:89-132.

[37] Menter F, Egorov Y. Scale-adaptive simulation method for unsteady flow predictions. Part 1 Theory and model description. Journal of Flow Turbulence and Combustion. 2010:85:113-138.

[38] Smagorinsky J. General circulation experiments with the primitive equations. Monthly Weather Reviews. 1963;91:99-165.

[39] Tong J, Abraham J, Tse J, Minkowycz W, Sparrow E. New archive of heat transfer coefficients from square and chamfered cylinders in cross flow. International Journal of Thermal Sciences. 2016;105:218-223.

[40] Tong J, Abraham J, Tse J, Sparrow E. Impact of chamfer contours to reduce column drag. Engineering and Computational Mechanics. 2015;168:79-88.

[41] S Bhattacharyya, H Chattopadhyay, AC Benim, Simulation of heat transfer enhancement in tube flow with twisted tape insert, Progress in Computational Fluid Dynamics, an International Journal, 2017; 17(3): 193-197.

[42] S Bhattacharyya, H Chattopadhyay, S Bandyopadhyay, S Roy, A Pal, S Bhattacharjee, Experimental investigation on heat transfer enhancement by swirl generators in a solar air heater duct, International Journal of Heat and Technology, 2016; 34(2): 191-196.

[43] M W Alam, S Bhattacharyya, B Souayeh, K Dey, F Hammami, M Rahimi-Gorji, R Biswas, CPU heat sink cooling by triangular shape micro-pin-fin: Numerical study, International Communications in Heat and Mass Transfer, 2020;112:104455

[44] S Bhattacharyya, H Chattopadhyay, Computational of studies of heat transfer enhancement in turbulent channel flow with twisted strip inserts, proceedings of CHT-15. 6th International symposium on advances in computational heat transfer, 2015: 209-220. DOI: 10.1615/ICHMT.2015. IntSympAdvComputHeatTransf.160

[45] S Bhattacharyya, H Chattopadhyay, R Biswas, DRE Ewim, Z Huan, Influence of inlet turbulence intensity on transport phenomenon of modified diamond cylinder: A numerical study, Arabian Journal for Science and Engineering, 2020; 45(2):1051-1058.

[46] S Bhattacharyya, S Das, A Sarkar, A Guin, A Mullick, Numerical simulation of flow and heat transfer around hexagonal cylinder, International Journal of Heat and Technology, 2017;35 (2):360-363

[47] Turbulence Modeling Resource (NASA). https://turbmodels.larc.nasa. gov

[48] ANSYS CFX Theory Guide, Published 2009, ANSYS, Inc.

Chapter 3

A New Forced Convection Heat Transfer Correlation for 2D Enclosures

Alejandro Rincón-Casado and
Francisco José Sánchez de la Flor

Abstract

This work presents a new parametric correlation for 2D enclosures with forced convection obtained from CFD simulation. The convective heat transfer coefficient of walls for enclosures depends on the geometry of the enclosure and the inlet and outlet openings, the velocity and the air to wall temperature difference. However, current correlations not dependent on the above parameters, especially the position of the inlet and outlet, or the temperature difference between the walls. In this work a new correlation of the average Nusselt number for each wall of the enclosure has been developed as a function of geometrical, hydrodynamic and thermal variables. These correlations have been obtained running a set of CFD simulations of a 3 m high sample enclosure with an inlet and outlet located at opposite walls. The varying parameters were: a) the aspect-ratio of the enclosure (L/H = 0.5 to 2), b) the size of the inlet and outlet (0.05 m to 2 m), c) the inlet and outlet relative height (0 m to 3 m high), and d) the Reynolds number ($Re_{in} = 10^3$ to 10^5). Furthermore, a parametric analysis has been performed changing the temperature boundary conditions at the internal wall and founds a novel correlation function that relates different temperatures at each wall. A specifically developed numerical model based on the SIMPLER algorithm is used for the solution of the Navier–Stokes equations. The realisable turbulence k-ε model, and an enhanced wall-function treatment have been used. The heat transfer rate results obtained are expressed through dimensionless correlation-equations. All developed correlations have been compared with CFD simulations test cases obtaining a $R^2 = 0.98$. This new correlation function could be used in building energy models to enhance accuracy of HVAC demands calculation and estimate the thermal load.

Keywords: Correlation heat transfer coefficient, Forced convection, CFD simulation, Square cavity, Rectangular enclosures

1. Introduction

The mechanisms for this energy transport are conduction, radiation and convection. While the study of conduction and radiation processes are founded on well-established analytical and numerical models, the treatment of convection is currently much less rigorous and precise (J. [1, 2]).

To characterise convection phenomena, fluid dynamics problems must be solved through computational fluid dynamics (CFD). The geometric complexity and variety of possible air-flow patterns and the integration of fluid dynamics problems with thermal simulation programs for buildings greatly complicate the treatment of these problems.

Heat-transfer coefficients inside of a building quantify the heat transfer rate between a wall and the surrounding air. Currently, thermal simulation programs assume that this coefficient has a constant value or can be calculated by assuming correlations for flat plates or using empirical correlations [3]. These hypotheses produce underestimations or overestimations of the heat-transfer coefficient and directly influence the air-conditioning demands for the building (J. [4]).

Heat-transfer coefficients are difficult to calculate because of the thermal and hydrodynamic boundary layer produced between the wall and fluid. Such calculations can be achieved through CFD tools and the use of dense meshes in close proximity to the walls. These procedures can capture temperature and velocity gradients, which are used to obtain the heat flow of the walls and heat-transfer coefficient. Because of the high computational complexity of these problems, heat-transfer coefficients for an entire building are calculated according to interconnected individual spaces.

However, the reference air temperature is a key factor in the calculation of heat-transfer coefficients according to Newton's law of cooling. According to the study conducted by Saena [5], significant discrepancies are found among reported results, with many authors calculating heat-transfer coefficients using the air temperature in the immediate proximity of the wall, air temperature 10 mm away from the wall, and fluid temperature at the inlet; however, the majority use the average temperature of the room.

Thus far, studies have not addressed the general problem of a two-dimensional (2D) room under forced convection in which all of the geometric and hydrodynamic parameters of the problem are varied. Studies have obtained correlations that depend on some but not all of the geometric parameters. For example, the work by Al-Sanea [6] analyses the fixed positions and sizes of the inlet and outlet openings and the influence of the aspect ratio of the room and Reynolds number on the heat-transfer ratio at the roof. In the case of ASHRAE [3], the room is treated as a flat plate, and the flow pattern within the enclosure, which is determined by its geometry and the positions and sizes of its openings, is ignored.

Another interesting study of 2D models was conducted by Saeidi [7], who produced a 2D model with a fixed position of the inlet and variable position of the outlet, the size of both openings and the Reynolds number. That study, however, is centred in the laminar flow regime, and the working fluid is water; in addition, the authors presented results for the local Nusselt number at the walls.

One of the first studies on ventilated inner enclosures was conducted by [8], who focused on 2D models of k-ε turbulence. The results of that work have been experimentally verified.

In the works by Novoselac et al. [9–11], the effects of discrete heat flows in the heat transfer by convection in enclosures with displacement ventilation systems were studied. Regarding the correlation proposed by the aforementioned authors regarding the floor geometry, a correction factor is used in the correlation (Tw-Tin)/(Tw-Ta) to calculate the heat-transfer coefficient in forced convection, and it is a function of the temperatures of the floor and air entering the room.

In addition, a correlation study conducted by Beausoleil-Morrison [1, 12] characterised flow regimes that are commonly found in real buildings, and the most appropriate correlations were selected from among all of the available correlations. Their goal was to create an adaptive algorithm for modelling heat transfer by convection in programs for thermal simulations in buildings. The most adequate correlation was proposed by Churchill and Usagi [13] in mixed convection using an exponent n = 3 and employing the natural convection correlations of Alamdari and Hammond [14] and corrections for forced convection by Fisher [15].

Recently, attempts have been made to couple CFD and building energy simulation (BES) techniques to obtain more precise results. Zhai and Chen [16] recommended an iterative coupling method that transfers the temperatures of the inner room surface from BES to CFD and returns the convection, heat-transfer coefficient, and air temperature in the room from CFD to BES.

J.M. Salmeron [17] studied the efficiency of cross ventilation at night as a function of the air flow rate, flow pattern and distribution of thermal mass and considered the positions of the inlet and outlet openings as well as their influence on the heat-transfer coefficients of the walls through the use of CFD techniques; however, the sensitivity was not analysed for all of the geometric variables of the problem in this study.

Sudhir [18] a numerical simulation is carried out to analyse the effect of turbulent intensity on the flow behaviour of flow past two dimensional bluff bodies. Triangular prism, diamond and trapezoidal shaped bodies with the same hydraulic diameter D, a dimensionless length scale are taken into consideration as bluff bodies. The study reveals that transition SST Model can be efficiently used to cover both laminar and turbulent flow regimes to estimate the heat transfer. However, this study does not show attention to the size of the inlets and outlets made in this study for rectangular enclosures.

All of the published research works thus far have been strongly centred on turbulence models in the type of code used to solve the Eqs. (Q. [19]) or to reproduce the experimental results [8]. However, only limited studies have considered all of the variables required to calculate heat-transfer coefficients inside of an enclosure and obtained correlations that can be easily implemented in applications for thermal simulations. The latter is the goal of this work, where the precision of the results is a function of the accuracy of the correlations.

Our aim is to obtain a series of correlations that calculate the convective heat-transfer coefficients at each wall of a 2D model under forced convection, with inlet and outlet openings in opposite walls. The most general solution of this problem assumes that the correlations depend on the airflow velocity upon entry, dimensions of the enclosure and positions and sizes of the inlet and outlet openings.

First, we studied cases in which all of the walls are at the same temperature to obtain a correlation that is independent of the temperatures of the walls or air. Then, we studied cases where the walls are at different temperatures to introduce a correction factor into the correlation that is dependent on the temperatures of the walls and air entering the enclosure.

2. Material and methods

2.1 Formulation

An enclosure with 2D geometry under forced convection have been studied. This model corresponds to enclosures where the inlet and outlet openings are as large as the third dimension. In these cases, the 3D effects of flow are negligible compared with the effects that occur in the 2D plane [5]. In a 2D enclosure under forced convection with only one inlet and one outlet, three distinct typologies occur: typology 1 includes an inlet and outlet that are located on opposite walls; typology 2 includes an inlet and outlet that are located on adjacent walls; and typology 3, includes an inlet and outlet that are located on the same wall. The present work focuses on typology 1, which is the most common in construction, although the developed methodology is valid for typologies 2 and 3 as well. **Figure 1** shows the geometric variables of the case being studied, including the dimensions of the enclosure, the positions and dimensions of the inlet and outlet openings and the temperature of the entry and walls.

For the formulation of the problem working with dimensionless numbers, the most important being the following:

$$Gr = \frac{g \cdot \beta \cdot (T_s - T_\infty)L_c^3}{\nu^2} \tag{1}$$

$$Pr = \frac{\nu \cdot \rho \cdot Cp}{k} \tag{2}$$

$$Ra_L = Gr \cdot Pr \tag{3}$$

$$Re_{in} = \frac{V_{in} \cdot W_{in}}{\nu} \tag{4}$$

$$Nu_i = \frac{h_i \cdot W_{in}}{k} \tag{5}$$

$$Ri = \frac{Gr}{Re^2} \tag{6}$$

$$y^+ = y \cdot \frac{u_\tau}{\nu} \tag{7}$$

The Richardson number indicates the relative importance of natural convection with respect to forced convection in mixed convection processes. This number determines the processes that are more important for convection: For low Richardson numbers (Ri < <1), the Reynolds number is larger than the Grashof number, and forced convection is predominant. In the opposite case (Ri > > 1), the effects of forced convection are negligible compared with that of natural convection. However, if the Richardson number lies somewhere between these two limits, both effects are important, and convection is considered mixed. The present work studies cases of predominant forced convection and does not consider the effects of natural convection.

2.2 CFD governing equations

To calculate the heat-transfer coefficients, the velocity, temperature and pressure fields of the enclosure must be determined. Thus, we employed the

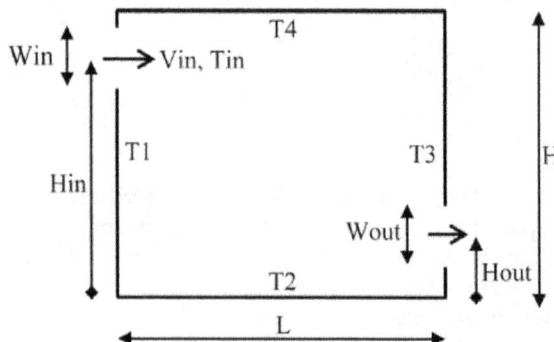

Figure 1.
Definition of the variables for enclosures with inlet and outlet openings on opposite walls.

Navier–Stokes equations, which describe the fluid motion for a given set of boundary conditions. These equations along with the turbulence model and energy equation are solved at each node of the mesh.

The turbulence model employed here is the realisable k-ε model. This model differs from the standard k-ε model through a new formulation of turbulent viscosity and transport equation for ε. The equations for the 3D model are provided in tensor notation, where x_i represents the variables X, Y and Z and u_i represents the corresponding velocity components.

The continuity equation is as follows:

$$\frac{\partial \rho}{\partial t} + \frac{\partial(\rho u_i)}{\partial x_i} = 0 \tag{8}$$

The equation for conservation of momentum is as follows:

$$\frac{\partial(\rho u_i)}{\partial t} + \frac{\partial(\rho u_i u_j)}{\partial x_j} = -\frac{\partial p}{\partial x_i} + \frac{\partial}{\partial x_j}\left[\mu\left(\frac{\partial u_i}{\partial x_j} + \frac{\partial u_j}{\partial x_i} - \frac{2}{3}\delta_{ij}\frac{\partial u_k}{\partial x_k}\right)\right] \tag{9}$$

The energy equation is as follows:

$$\frac{\partial(\rho T)}{\partial t} + \frac{\partial(\rho u_i T)}{\partial x_i} = \frac{\partial}{\partial x_i}\left(\left(\frac{\mu}{Pr} + \frac{\mu_t}{Pr_T}\right)\frac{\partial T}{\partial x_i}\right) \tag{10}$$

The shear viscosity equation is as follows:

$$\mu_t = \rho C_\mu k^2 / \varepsilon \tag{11}$$

The variable C_μ is calculated as follows:

$$C_\mu = \frac{1}{4.04 + A_s \frac{kU^*}{\varepsilon}} \tag{12}$$

where

$$U^* \equiv \sqrt{S_{ij}S_{ij} + \tilde{\Omega}_{ij}\tilde{\Omega}_{ij}}; \ \tilde{\Omega}_{ij} = \Omega_{ij} - 2\varepsilon_{ijk}\omega_k; \ A_s = \sqrt{6}\cos\phi; \ S_{ij} = \frac{1}{2}\left(\frac{\partial u_j}{\partial x_i} + \frac{\partial u_i}{\partial x_j}\right) \tag{13}$$

The "k" transport equation in the turbulence model is as follows:

$$\frac{\partial(\rho k)}{\partial t} + \frac{\partial(\rho k u_j)}{\partial x_j} = \frac{\partial}{\partial x_j}\left[\left(\mu + \frac{\mu_T}{\sigma_k}\right)\frac{\partial k}{\partial x_j}\right] + G_k + G_b - \rho\varepsilon + S_k \tag{14}$$

where G_k represents the production of turbulent kinetic energy, which is common to all k-ε turbulence models and is given by

$$G_k = -\overline{u'_i u'_j}\frac{\partial u_j}{\partial x_i} \tag{15}$$

The term G_b represents the generation of turbulent kinetic energy because of buoyant forces when the system is under a gravitational field, and it is calculated as follows:

$$G_b = \beta g_i \frac{\mu_T}{Pr_t} \frac{\partial T}{\partial x_i} \qquad (16)$$

where $Pr_t = 0.72$ is the Prandtl number for energy and β is the thermal expansion coefficient, which is calculated as follows:

$$\beta = -\frac{1}{\rho}\left(\frac{\partial \rho}{\partial x_i}\right) \qquad (17)$$

The transport equation for ε from the turbulence model is as follows:

$$\frac{\partial(\varepsilon)}{\partial t} + \frac{\partial(\rho \varepsilon u_j)}{\partial x_j} = \frac{\partial}{\partial x_j}\left[\left(\mu + \frac{\mu_T}{\sigma_\varepsilon}\right)\frac{\partial \varepsilon}{\partial x_j}\right] + \rho C_1 S\varepsilon - \rho C_2 \frac{\varepsilon^2}{k + \sqrt{\nu\varepsilon}} + C_{1\varepsilon}\frac{\varepsilon}{k}C_{3\varepsilon}G_b + S_\varepsilon \qquad (18)$$

The source terms S_ε and S_k can be defined for each case and are optional. The coefficient $C_{3\varepsilon}$ is calculated as follows:

$$C_{3\varepsilon} = \tanh\left|\frac{v}{u}\right| \qquad (19)$$

The constants used in the realisable k- ε model are as follows:

$$\sigma_k = 1.0, \sigma_\varepsilon = 1.3, C_{1\varepsilon} = 1.44, C_{2\varepsilon} = 1.92, C_2 = 0.43, C_2 = 1.9 \qquad (20)$$

The heat-transfer coefficient between the surface of the wall and fluid in motion at different temperatures is provided by Newton's law of cooling and is dependent on the total heat of the wall, the transfer area and the temperature difference between the surface and unperturbed fluid:

$$\bar{h}_i = \frac{Q_i}{A_i(T_{si} - T_\infty)} \qquad (21)$$

The total heat flow that is transferred between the fluid and the wall is calculated from the temperature gradient produced inside of the thermal boundary layer of the fluid through an integration of Fourier's law at each wall. Because the heat flow depends on the temperature gradient, it is solved using CFD. Thus, the heat flow at a wall is obtained by integrating the temperature gradient along the wall, and it is affected by the fluid's conductivity, as shown in Eqs. (22) and (23):

$$Q_{CFD}\Big|X = \int_0^L k \frac{\partial T}{\partial y} dx \Big|_{y=0} \qquad (22)$$

$$Q_{CFD}\Big|Y = \int_0^H k \frac{\partial T}{\partial x} dy \Big|_{x=0} \qquad (23)$$

Therefore, at the fluid–solid interface, the heat transfer by conduction equals the heat transfer by convection, and the average heat-transfer coefficients are as follows:

$$\bar{h}_{ix} = \frac{\int_0^L k \frac{\partial T}{\partial y} dx \Big|_{y=0}}{A_i(T_{si} - T_\infty)} \qquad (24)$$

$$\overline{h}_{iy} = \frac{\int_0^H k \frac{\partial T}{\partial x} dy \Big|_{x=0}}{A_i(T_{si} - T_\infty)} \tag{25}$$

Eqs. (24) and (25) show that the temperature difference between the wall (T_{si}) and free fluid (T_∞) is proportional to the temperature gradient of the thermal boundary layer. Thus, increases of the temperature transferred between the wall and fluid correspond to increases of the gradient within the boundary layer, which maintains a constant ratio and indicates that the heat-transfer coefficient is temperature independent. This finding is valid for flat plates where the fluid is not perturbed by other walls and for enclosures with walls whose temperatures are all equal. However, this finding is not valid for cases in which the walls have different temperatures.

2.3 Computational model

The computational model presented here considers a steady flow, 2D geometry and incompressible Newtonian fluid. All of the fluid properties remain constant, and it behaves as an ideal gas. All of the properties are evaluated at the fluid's average temperature within the enclosure. The phenomenon under study is forced convection; therefore, the velocities are large enough for all buoyancy effects to be negligible and for gravity to be ignored. The CFD results are obtained by solving the Navier–Stokes equations and energy equation through the finite volume method using the commercial software package Fluent V14 [20]. The SIMPLE (Semi-Implicit Method for Pressure Linked Equations) numerical algorithm developed by Patankar and Spalding [21] is also used. The equations for mass, momentum and energy are solved iteratively for the corresponding boundary conditions using numerical methods until convergence is reached for the variables of interest (velocity, temperature, pressure and heat flow at the walls).

One of the most important aspects to consider when using CFD tools is the construction of the mesh for the computational environment. The mesh utilised in the simulation is built from rectangular elements, and it enlarges from the walls toward the centre of the enclosure in a uniform fashion. This type of mesh produces a better convergence because there is a higher density of elements within the thermal boundary layer where the heat-transfer coefficient is calculated. To ensure the correct solution within the boundary layer, nodes must be placed within the viscous sub-layer. Thus, the first element must be located at a maximum distance of 1 mm from the wall, and the growth rate toward the centre of the enclosure must be between 10% and 15% [22].

The parameter that controls the correct solution of the viscous sub-layer is y+. This dimensionless parameter depends on the turbulence model [Eq. (26)]. Thus, for k-ε turbulence with "enhanced wall treatment," the parameter y + must have a value of approximately one. **Figure 2** shows an example of an enclosure represented by a mesh with an aspect ratio of one.

$$y^+ = y \cdot \frac{u_\tau}{\nu} \tag{26}$$

2.4 Validation of the CFD methodology

To validate the CFD methodology employed in 2D enclosures under forced convection, the following problem with a known solution is used as a reference: the case of a flat plate. By solving this problem using the CFD method and comparing

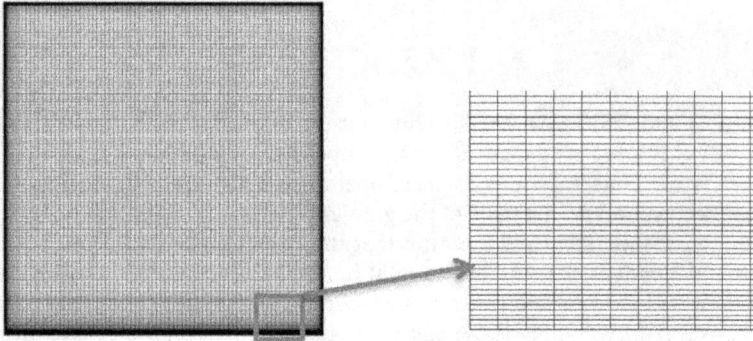

Figure 2.
Uniform-growth mesh for a 2D enclosure under forced convection.

the results with those obtained by correlating the flat plate in both the laminar and turbulent regimes, the CFD methodology is validated. The problem to be solved is that of a horizontal plate of length L with an incident air flow at velocity v_{in} in the parallel direction. The circulating air is at temperature T_∞, whereas the flat plate is at temperature T_s. The computational domain is large enough so that it does not influence the solution. It is necessary to ensure that the number of nodes and type of mesh do not depend on the solution. The correct solution depends on the parameter y+, which must have a value of approximately 1. Therefore, the number of mesh elements and elements located within the limiting thermal layer for each case are adjusted to obtain y + $\approx$1.

To generalise the studied cases, the Reynolds number is set between 3 x·10^3 and 7 x·10^6 along the length of the plate in the direction of the air velocity. Thus, both the laminar and turbulent regimes are explored. The studied cases are presented in **Table 1**. To validate the methodology for the case of a flat plate, the CFD results are represented in terms of the average Nusselt number along with the results obtained through the flat plate correlations by Pohlhausen [23] and Reynolds analogy. As shown in **Figure 3**, the match between the CFD results and correlations is good and presents a relative error of less than 3%. Thus, the methodology employed for the mesh, convergence criterion and turbulence model are correct and can be used in the solution of 2D enclosures under forced convection.

2.5 Case studies

2.5.1 Enclosures with walls at the same temperature

In 2D enclosures under forced convection where all of the walls are at the same temperature, which is generally different from the air temperature at entry, the variables can be varied continuously. For the present work, only three values are considered for each variable. The typical enclosure height employed for construction is 3 m; therefore, this variable is kept fixed and used to rescale all other variables and make them dimensionless. **Table 1** shows the values used for each dimensionless variable. To solve the problem, a factorial experiment is conducted with six factors and four responses [24]. The variables are labelled as (X_j), and the variable of interest is labelled as a response Y_i. The relationship between factor and response is provided by the function $Y_i = f(X_i)$.

For the 2D case being studied, there are four responses: the average heat-transfer coefficients at each wall (Nu_1, Nu_2, Nu_3 and Nu_4) and six factors or independent variables (V_{in}/H, L/H, W_{in}/H, W_{out}/H, H_{in}/H and H_{out}/H). **Table 2** shows the

Case	L(m)	V(m/s)	Re	Regime	Mesh	$\Delta(1^{st}node)$	yplus	Nu
1	1	0.5	3.31E+04	Laminar	300Vx100H	0.001	—	56.1
2	1	1	6.63E+04	Laminar	300Vx100H	0.001	—	78.3
3	1	2	1.33E+05	Laminar	300Vx100H	0.001	—	109.7
4	1	3	1.99E+05	Laminar	300Vx100H	0.001	—	133.8
5	1	4	2.65E+05	Laminar	300Vx100H	0.001	—	154.2
6	2	0.1	1.33E+04	Laminar	300Vx200H	0.001	—	36.5
7	2	0.5	6.63E+04	Laminar	300Vx200H	0.001	—	79.0
8	2	1	1.33E+05	Laminar	300Vx200H	0.001	—	110.5
9	2	1.5	1.99E+05	Laminar	300Vx200H	0.001	—	134.6
10	2	2	2.65E+05	Laminar	300Vx200H	0.001	—	154.8
11	2	3	3.98E+05	Laminar	300Vx200H	0.0001	—	188.8
12	3	3.5	6.96E+05	Turbulent	300Vx300H	0.0001	0.63	880.1
13	3	4	7.95E+05	Turbulent	300Vx300H	0.0001	0.71	981.5
14	3	4.5	8.95E+05	Turbulent	300Vx300H	0.0001	0.79	1082.5
15	3	5	9.94E+05	Turbulent	300Vx300H	0.0001	0.87	1181.8
16	3	6	1.19E+06	Turbulent	300Vx300H	0.0001	1.03	1375.9
17	6	7	2.78E+06	Turbulent	300Vx600H	0.0001	1.11	2722.1
18	6	10	3.98E+06	Turbulent	300Vx600H	0.0001	1.53	3661.3
19	6	15	5.96E+06	Turbulent	300Vx600H	0.0001	1.21	5161.7
20	6	17	6.76E+06	Turbulent	300Vx600H	0.0001	1.48	5754.2

Table 1.
Validation cases for a flat plate and forced convection.

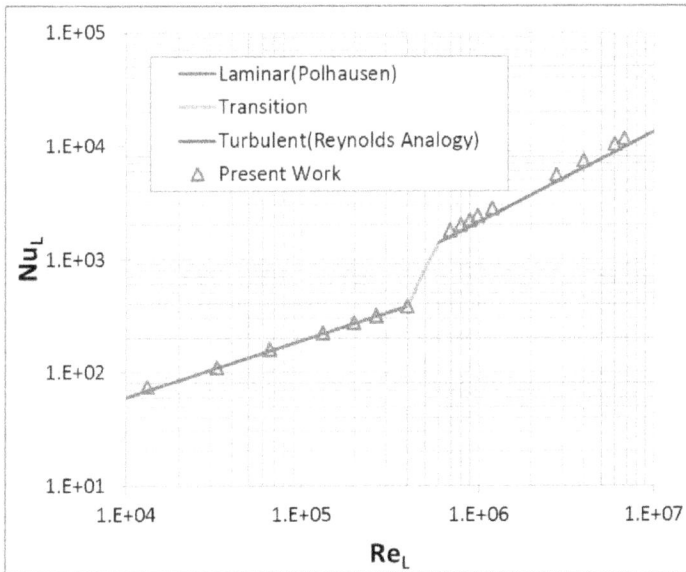

Figure 3.
Validation of the CFD methodology for a flat plate and forced convection.

Vars. Adim.	Min.	Med.	Max.
W_{in}/H	0.017	0.067	0.4
W_{out}/H	0.017	0.067	0.4
H_{in}/H	0.025-0.083-0.2166	0.5	0.975-0.917-0.783
H_{out}/H	0.025-0.083-0.2166	0.5	0.975-0.917-0.783
L/H	0.5	1	2
Re_{in} (v = 0.5)	1657	6627	39764
Re_{in} (v = 1)	3314	13255	79529
Re_{in} (v = 1.5)	4971	19882	119293

Table 2.
Range of application of the variables (dimensionless).

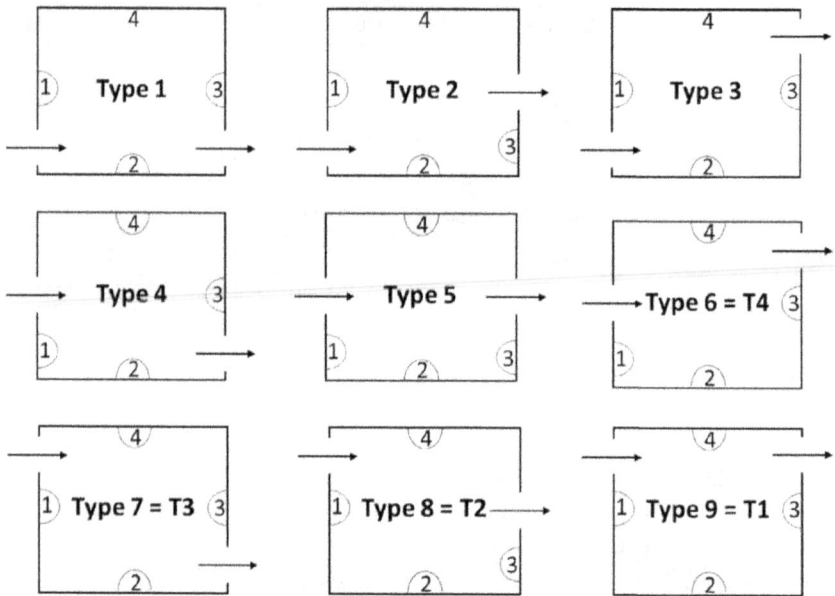

Figure 4.
Study cases for the positions of the inlet and outlet openings for typology 1.

numerical value of each variable.. The goal is to determine the function f that best fits the results from the simulation, and it is therefore necessary to solve the full range of cases to obtain the most general correlation possible.

To obtain enough data and the best fit to the results, all of the variables must be combined with their three possible values so that any possible combination within the variable applicability range is contemplated. Thus, the six variables can take three values each, which produces a total of $3^6 = 729$ cases. However, because the mechanism being studied is forced convection, gravity does not influence the motion of the fluid, which means that the fluid motion is independent of the spatial orientation of the enclosure. Therefore, we can infer that among all possible cases, certain cases will be symmetric with respect to the positions of the inlet and outlet (H_{in} and H_{out}). **Figure 4** shows all of the possible cases when only the positions of the inlet and outlet are varied.

Figure 4 shows that there are nine possible typologies as a function of the positions of the inlet and outlet openings. It is only necessary to study five out of these nine because the remaining typologies are symmetric cases because of forced convection. In these symmetric cases, the heat-transfer coefficients at the inlet and outlet (walls one and three, respectively) are equal; the coefficient of wall two is equal to that of wall four in its symmetric case, and wall four is equal to that of wall two in its symmetric case. Therefore, symmetry considerations reduce the number of cases to be studied from nine to five, thus affecting the values used for the variables H_{in} and H_{out}. The corresponding number of cases is reduced from 729 down to 405 (3·x 3·x 3·x 3·x 5). To avoid convergence problems and inconsistent results, all of the cases where the inlet opening is much larger than the outlet opening ($W_{in} >> W_{out}$) are eliminated. Thus, out of the nine possible combinations of W_{in} and W_{out}, three are eliminated. This reduced the number of cases from 405 to 270 (3·x 3·x 6·x 5).

Through this analysis, we have reduced the number of study cases as well as the simulation time and work involved in elaborating the meshes. Therefore, the parametric study is feasible, with 270 cases to be solved, and statistical methods are not required to reduce the range of study cases.

2.5.2 Enclosures with walls at different temperatures

In the cases where the wall temperatures are all different, the heat-transfer coefficient depends on the temperature distribution of the enclosure walls regardless of the occurrence of forced convection. **Table 3** shows a comparison of the results for walls at equal temperatures with those for walls at different temperatures to demonstrate this influence.

Table 3 shows that the heat-transfer coefficients obtained for different wall temperatures do not match those for walls at equal temperatures because the fluid temperature considered in Newton's law of cooling is the average enclosure temperature (T_a), which is different from that of unperturbed fluid (T_∞). In cases where the wall temperatures are equal, both temperatures (T_a and T_∞) are consistent, and the heat-transfer coefficients are equivalent for all cases. However, when the walls temperatures are not equivalent, the mean temperature of enclosure T_a is not consistent with the unperturbed fluid temperature T_∞, which leads to different heat-transfer coefficients.

Case	Tin	T1	T2	T3	T4	Ta	T∞	h1	h2	h3	h4
1	20	30	30	30	20	23.28	23.3	2.08	2.41	2.92	5.10
2	20	40	40	40	30	26.56	26.6	2.08	2.41	2.92	5.10
3	20	50	50	50	20	46.25	46.2	2.08	2.41	2.92	5.10
4	40	30	30	30	20	36.72	36.7	2.08	2.41	2.92	5.10
5	30	20	20	50	20	45.62	27.0	0.59	0.76	16.39	1.46
6	30	40	50	20	10	28.74	31.7	1.14	2.10	4.17	2.30
7	20	50	30	50	40	38.83	28.5	4.64	0.23	5.63	9.36
8	50	10	10	20	10	33.27	39.0	2.55	2.96	4.13	7.51
9	40	40	10	10	10	48.65	33.6	2.54	1.57	1.86	2.49
10	50	20	50	10	30	43.87	40.1	2.23	4.13	2.61	4.21

Table 3.
Casos estudiados en 2D con convección forzada y temperaturas de las Paredes distintas.

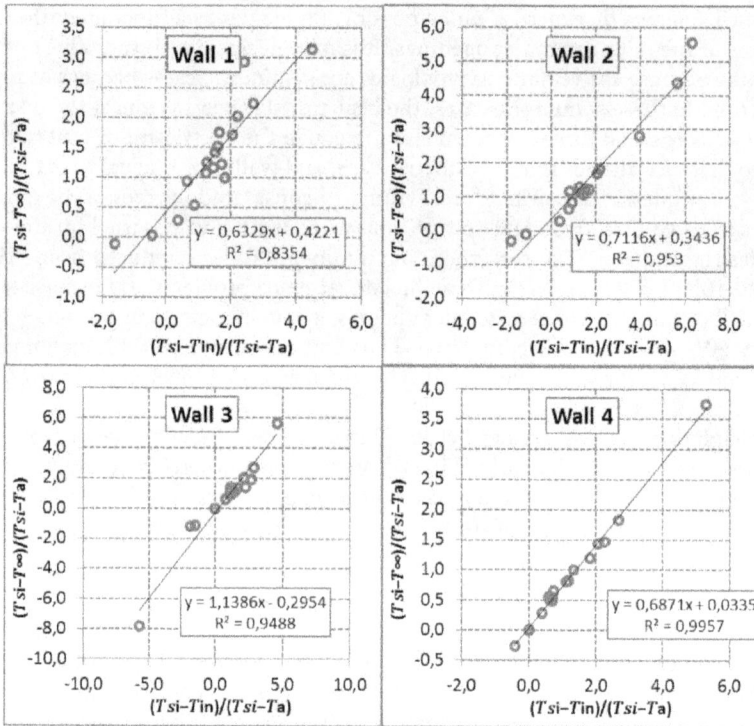

Figure 5.
Relationship between T_{in} and $T_{\infty i}$ for each wall.

In these cases, a correction factor for temperature must be obtained that considers this phenomenon. This factor (F_i) relates the heat-transfer coefficient in the case of different wall temperatures ($h^{T_i \neq}$) to the corresponding value for equal wall temperatures (h^{corr}) as follows:

$$h_i^{T_i \neq} = h_i^{corr} \cdot F_i \tag{27}$$

Therefore, for a given $T_{\infty i}$ for cases solved using CFD, a correcting factor for the temperature jump ($Ts_i - T_a$) must be obtained so that the heat-transfer coefficient and the average air temperature T_a can be calculated from this correcting factor. Newton's law of cooling and the heat-transfer coefficient h_i^{corr} for the case of walls with equal temperatures are applied, and the correcting factor is provided by Eq. (30):

$$Q_{wall\_i}^{CFD} = A_i h_i^{corr} (Ts_i - T_{\infty i}) \tag{28}$$

$$Q_{wall\_i}^{CFD} = A_i h_i^{corr} (Ts_i - T_a) \cdot F_i \tag{29}$$

$$F_i = \frac{(Ts_i - T_{\infty i})}{(Ts_i - T_a)} \tag{30}$$

Eq. (30) shows that the correction factor F_i depends on $T_{\infty i}$, which is calculated using CFD through Newton's cooling law as well as through the heat flow and heat-transfer coefficient obtained from the following correlation:

$$T_{\infty i} = \frac{Q_{wall\_i}^{CFD}}{A_i h_i^{corr}} \tag{31}$$

However, the heat flow is only known for CFD cases; therefore, a correction factor must be calculated that does not depend on T_∞ but depends on a known variable, such as T_{in}. This temperature was employed by Novoselac [10, 11] to obtain the correction factor in his experiments. **Figure 5** shows the temperature difference using T_{in} and $T_{\infty i}$, which produces an excellent match; thus, the variable T_{in} can be used in the correction factor.

We conclude that the correction factor can be written as a function of T_{in}, T_a and T_{si} in the following equation:

$$F_i = \frac{Ts_i - T_{\infty i}}{Ts_i - T_a} = f\left(\frac{Ts_i - T_{in}}{Ts_i - T_a}\right) \tag{32}$$

This correcting factor depends on the flux and positions of the inlet and outlet openings. Thus, we have demonstrated how to calculate this factor and must determine a correlation expression for the factor for each typology of the enclosure. For that purpose, 100 cases are simulated using CFD, with 20 cases for each of the five enclosure typologies where the temperatures of the wall and inlet are varied randomly. Thus, a correlation of the correction factor can be obtained for each typology and for each wall of the enclosure.

3. Result and discussion

All of the cases defined in the previous section are solved using CFD by calculating the heat flow at every wall and the average temperature of the air in the enclosure. The solutions to all of the cases are considered to have converged when the convergence criteria adopted in the CFD simulations are met. In our case, the criteria are two-fold: the residues of the equations must be under 10^{-8}; however, the monitored variable (heat flow at the wall) must converge to a constant value within at least 1000 iterations. With these constraints in the simulation software, most of the cases converge within 5000 iterations.

3.1 Proposed correlations for enclosures with walls at equal temperatures

To obtain a correlation that calculates the average heat-transfer coefficients of walls of a 2D enclosure based on geometric and hydrodynamic parameters, a mathematical treatment of the data is necessary to determine the function that best fits the simulation results. Additionally, the correlation must have the same form as the flat plate under forced convection. That is, it must depend on the following: the Reynolds number at the inlet to the power n; the Prandtl number for air (0.72); and a constant C that depends on all of the geometric parameters.

$$Nu_i = C \cdot Re_{in}^n Pr \tag{33}$$

$$C = f\left(\frac{L}{H}, \frac{W_{in}}{H}, \frac{W_{out}}{H}, \frac{H_{in}}{H}, \frac{H_{out}}{H}\right) \tag{34}$$

Once the simulation results are obtained, mathematical optimisation techniques are applied to obtain the correlation coefficients that best fit the results from the CFD simulations. The form of the correlation that best fits the CFD results is

Eq. (35), which is obtained in dimensionless form through the Nusselt number and Eq. (36). The correlation coefficients for the walls are shown in **Table 4**.

$$Nu_i = \left[a\frac{W_{in}}{H} + b\frac{W_{out}}{H} + c\frac{H_{in}}{H} + d\frac{H_{out}}{H} + e\frac{L}{H} + f \right] \cdot Re_{in}^{n} \tag{35}$$

$$h_i = \frac{Nu_i k}{W_{in}} \tag{36}$$

Figures 6 and 7 show a comparison between the results from correlations and those obtained from the simulations. A mean error of approximately 15% was observed; however, the correlation describes the phenomenon with sufficient precision despite the complexity of the cases being studied.

3.2 Proposed correlations for enclosures with walls at different temperatures

In cases where the walls are at different temperatures, the heat-transfer coefficient is obtained from the corresponding coefficient through correlations for walls at the same temperature with the correcting factor. This factor depends on the flow and the positions of the inlet and outlet. **Figure 8** shows a correlation of the correction factors for each wall of an enclosure of typology 2, where a variety of cases have been simulated, and the temperatures of the walls and air at the entry are

2D Correlation coefficients forced convection					
Wall		**Wall 1**	**Wall 2**	**Wall 3**	**Wall 4**
Coefficients	a	0.019512	0.041169	0.066311	0.009220
	b	−0.006984	−0.016679	−0.027515	−0.004926
	c	−0.005944	−0.033410	0.005247	−0.002107
	d	0.003044	−0.024028	0.005527	0.006855
	e	−0.005913	−0.002876	−0.003623	−0.001066
	f	0.016337	0.041022	0.020702	0.003956
	n	0.770041	0.782311	0.771561	0.868345

Table 4.
Correlation coefficients for 2D enclosures under forced convection.

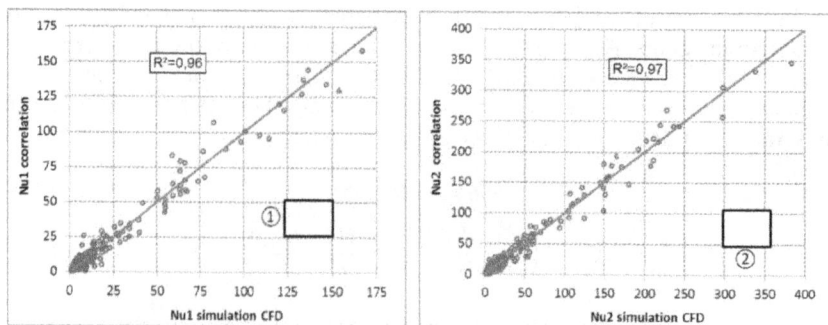

Figure 6.
Comparison of Nu1 and Nu2 obtained through correlations and simulations.

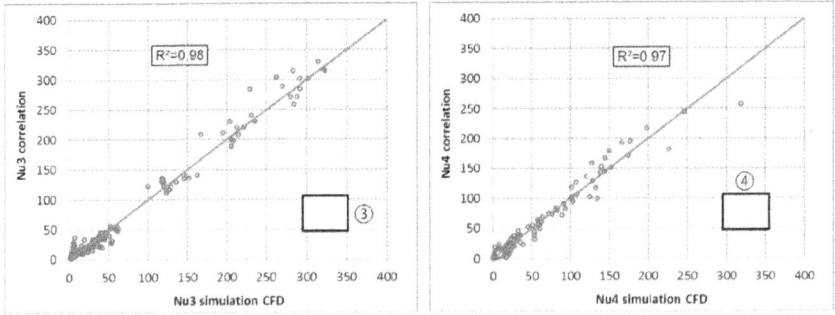

Figure 7.
Comparison of Nu3 and Nu4 obtained through correlations and simulations.

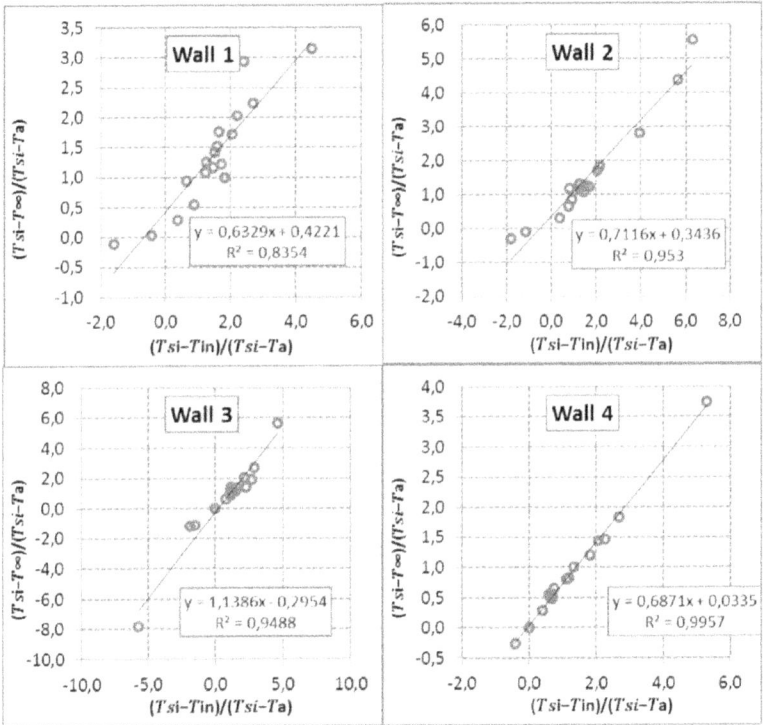

Figure 8.
Correlation between the correcting factor with $T_{\infty i}$ and T_{in}. Enclosure of typology 2.

varied. As shown in **Figure 8**, the fit and function that relates the correcting factor obtained through $T_{\infty i}$ to that obtained through T_{in} are good. To appreciate the quality of the fit, **Figure 9** shows the heat-transfer coefficient predicted using the correcting factor from CFD for each of the walls. The precision is high, which demonstrates that this factor must be employed to precisely calculate the heat-transfer coefficients when the walls have different temperatures. We proceed in the same fashion for the remaining typologies. **Table 5** shows the coefficients of the correlation for the correction factor for each wall and for the five typologies. Eq. (37) provides the heat-transfer coefficient when the wall temperatures are different.

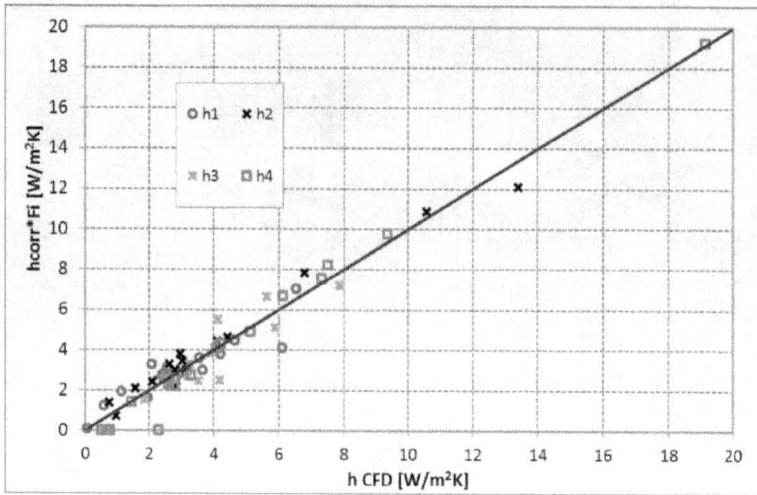

Figure 9.
Heat-transfer coefficient calculated using the correcting factor compared with the heat-transfer coefficient calculated using CFD.

Correlation of the correction factor F_i

$$F_i = a_i \left(\frac{T_{s_i} - T_{in}}{T_{s_i} - T_a} \right) + b_i$$

Type	Wall	a_i	b_i
Type 1	1	0.5672	0.2447
	2	0.1909	0.6831
	3	0	1
	4	0.4606	0.228
Type 2	1	0.6822	0.3319
	2	0.748	0.2974
	3	0.915	−0.1822
	4	0.7111	0.0077
Type 3	1	−0.4856	1.6965
	2	0.5175	0.2233
	3	0.4242	0.5216
	4	−0.1689	1.2546
Type 4	1	0	1
	2	0.2365	0.6331
	3	0.0000	1.0000
	4	−0.4644	1.6929
Type 5	1	0.5720	0.3993
	2	0.4951	0.7903
	3	0.7987	−0.1273
	4	0.4951	0.7903

Table 5.
Correcting factors for all typologies.

$$h_i^{T_i \neq} = h_i^{corr} \cdot \left(a_i \left(\frac{Ts_i - T_{in}}{Ts_i - T_a} \right) + b_i \right) \tag{37}$$

3.3 Effect of the ACH and W_{in} on the heat-transfer coefficients

To analyse the effect of the number of air changes per hour on the heat-transfer coefficients for each of the typologies, we employ the correlation obtained for cases where the walls are at the same temperature. Within the application range of the

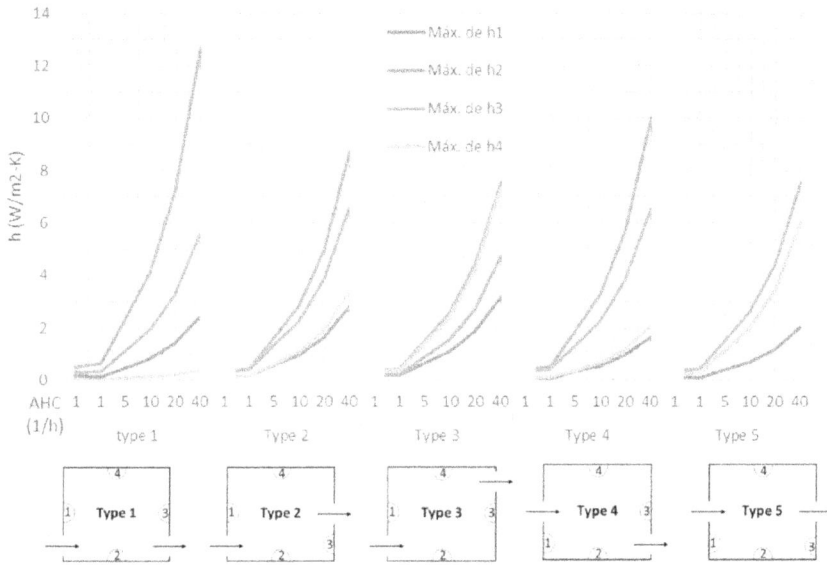

Figure 10.
Effect of the number of air changes per hour and typology on the heat-transfer coefficients at every wall.

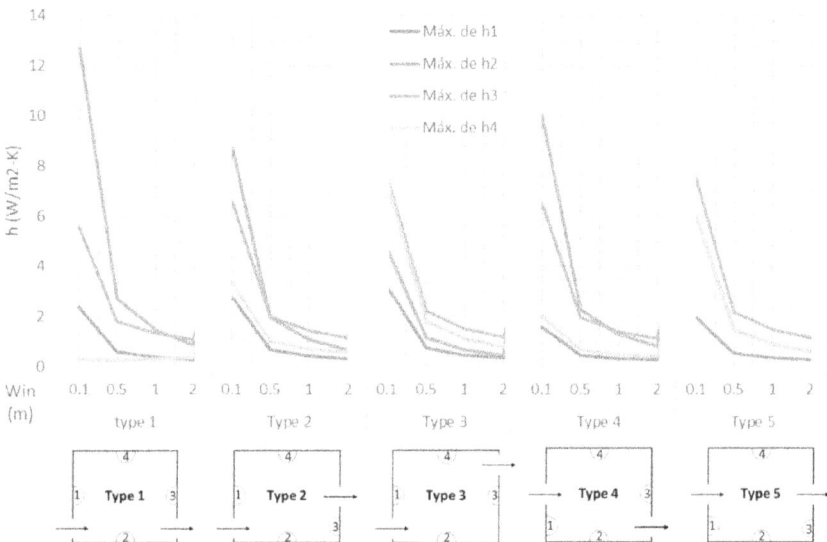

Figure 11.
Effect of the size of the inlet opening and typology on the heat-transfer coefficients at every wall.

variables, **Figure 10** shows the maximum values of the heat-transfer coefficients for each wall and typology compared with the number of air changes per hour. The maximum values are reached at wall two and its symmetric pair wall four; however, wall one shows the least sensitivity to variations in the number of changes. In cases where the enclosure is used for storing energy during the day and releasing it at night through ventilation, **Figure 10** shows that typology 1 provides the highest heat transfer rates through cross ventilation. More specifically, the floor and roof (in the symmetric case) are the recommended enclosures for energy storage because they have higher heat-transfer coefficients and larger solid-to-fluid heat transfers compared with the rest of the enclosures. This is a good example that can be used as a design tool.

To analyse the influence of the variable W_{in} on the heat-transfer coefficients as a function of the typology, **Figure 11** shows the maximum values of the heat-transfer coefficients reached for each wall. The maximum values are obtained for typology 1, and in general, the coefficient decreases as the opening size increases because increases in opening size lead to decreases in the speed of the air that interacts with the walls and thus reduces the heat exchange between the wall and air.

4. Conclusions

Heat-transfer coefficients are important parameters in the thermal modelling of a building and significantly influence the required air conditioning. The complexity of calculating these coefficients requires the use of CFD techniques that are not available to all users, which justifies the calculation of correlations that depend on the geometric parameters of a 2D enclosure under forced convection. We found that the CFD results are similar to those obtained using correlations for the flat plate model. The latter has the advantage of being velocity-independent in the proximity of the walls and is only dependent on the air speed at the inlet, which is known.

Another aspect highlighted in this study is the calculation of a correction factor that can be used when the wall temperatures are different. In addition, we obtained a correlation of the correction factor for each topology, which correct for the heat-transfer coefficients when all of the walls are at the same temperature.

Thus, our results indicate that the variables with the highest influence on the heat-transfer coefficients are the number of air changes in an enclosure followed by the size of the inlet.

Acknowledgements

This work is linked to two research projects, the first, and most socialising, is the project of National Plan "The air.es" entitled "Numerical modeling of combined heat inside buildings aeraulics oriented design eco-efficient "(MTM2012-36124-C02-00). This project is developed in collaboration with the research group FQM120 "Mathematical Modelling and Simulation of Environmental Systems", University of Sevilla.

Nomenclature

H	enclosure height (m)
L	enclosure length (m)
Win	inlet opening dimension (m)

$Wout$	outlet opening dimension (m)
Hin	height inlet(m)
$Hout$	height outlet(m)
Tin	air inlet temperatura (°C)
Ti	wall temperature i (1, 2, 3, y 4)
Gr	Grashof number
Nu	Average Nusselt number
Pr	Prandtl number
Re	Reynolds number
Ri	Richardson number
Ra	Rayleigh number
Pr	Prandtl number
k	Thermal conductivity of air (W/mK)
ρ	density (kg/m3)
Cp	specific heat (J/kg·K)
ν	kinematic viscosity (m2/s)
β	coefficient of thermal expansion (1/K).
$y+$	y plus
μ_τ	friction velocity

Author details

Alejandro Rincón-Casado[1]* and Francisco José Sánchez de la Flor[2]

1 Escuela Superior de Ingeniería, Departamento de Ingeniería Mecánica y Diseño Industrial, Universidad de Cádiz, Puerto Real, Cádiz, Spain

2 Escuela Superior de Ingeniería, Departamento de Máquinas y Motores Térmicos, Universidad de Cádiz, Puerto Real, Spain

*Address all correspondence to: alejandro.rincon@uca.es

IntechOpen

References

[1] Beausoleil-Morrison I., An algorithm for calculating convection coefficients for internal building surfaces for the case of mixed flow in rooms, Energy and Buildings 33 (2001) 351–361.

[2] J. Clarke, Energy Simulation in Building Design, Butterworth-Heinemann, Oxford, 2001 (ISBN 0 7506 5082 6).

[3] Beausoleil-Morrison Peeters L., I., A. Novoselac, Internal convective heat transfer modeling: Critical review and discussion of experimentally derived correlations, Energy and Buildings, Volume 43, Issue 9, September 2011, Pages 2227-2239, ISSN 0378-7788

[4] J. Clarke, 1991, Internal convective heat transfer coefficients: a sensitivity study, Report to ETSU, Glasgow, UK.

[5] Al-Sanea Sami A., M.F. Zedan, M.B. Al-Harbi, Effect of supply Reynolds number and room aspect ratio on flow and ceiling heat-transfer coefficient for mixing ventilation, International Journal of Thermal Sciences, Volume 54, April 2012, Pages 176-187, ISSN 1290-0729, 10.1016/j.ijthermalsci.2011.12.007.

[6] ASHRAE, Handbook of Fundamentals, American Society of Heating, Refrigerating and Air-Conditioning Engineers, Inc., Atlanta, 1997.

[7] Saeidi S.M., J.M. Khodadadi, Forced convection in a square cavity with inlet and outlet ports, International Journal of Heat and Mass Transfer, Volume 49, Issues 11–12, June 2006, Pages 1896-1906, ISSN 0017-9310

[8] Nielsen P.V., A. Restivo, J.H. Whitelaw, The velocity characteristics of ventilated rooms, ASME Journal of Fluids Engineering 100 (1978) 291e298.

[9] Novoselac A., (2005), Combined energy and airflow simulation program

for building mechanical system design, Ph.D. Dissertation, Pennsylvania State University, USA.

[10] Novoselac A., B. Burley, J. Srebric (2006a), Development of new and validation of existing convection correlations for rooms with displacement ventilation systems, Energy and Buildings 38 (2006) 163–173.

[11] Novoselac A., B. Burley, J. Srebric (2006b), Development of new and validation of existing convection correlations for rooms with displacement ventilation systems, Energy and Buildings 38 (2006) 163–173.

[12] Beausoleil-Morrison, 2000, The adaptive coupling of heat and air flow modelling within dynamic whole-building simulation, Ph.D. Thesis, University of Strathclyde, Glasgow, UK.

[13] Churchill S., R. Usagi, A general expression for the correlation of rates of transfer and other phenomena, AIChE Journal 18 (1972) 1121–1128.

[14] Alamdari F., G. Hammond, Improved data correlations for buoyancy-driven convection in rooms, Building Services Engineering research and Technology 4 (1983) 106–111.

[15] Fisher D., 1995, An experimental investigation of mixed convection heat transfer in a rectangular enclosure, Ph. D. Thesis, University of Illinois, Urbana, USA.

[16] Zhai Zhiqiang, Qingyan Chen; *, Philip Haves, Joseph H. Klems. On approaches to couple energy simulation and computational fluid dynamics programs. Building and Environment 37 (2002) 857 – 864

[17] Jose Manuel Salmerón Lissen, Juan Antonio Sanz Fernández, Francisco José Sánchez de la Flor, Servando Álvarez

Domínguez, and Álvaro Ruiz Pardo
(2007) Flow Pattern Effects on Night
Cooling Ventilation. International
Journal of Ventilation: June 2007, Vol. 6,
No. 1, pp. 21-30.

[18] Sudhir Chandra Murmu, Suvanjan
Bhattacharyya, Himadri
Chattopadhyay, Ranjib Biswas, Analysis
of heat transfer around bluff bodies
with variable inlet turbulent intensity: A
numerical simulation, International
Communications in Heat and Mass
Transfer, Volume 117, 2020, 104779,
ISSN 0735-1933,

[19] Q. Chen, Comparison of different k-
e 3 models for indoor air flow
computations, Numerical Heat Transfer
28 (1995) 353e369 Part B.

[20] Fluent v14. Ansys inc: 2012.

[21] Patankar S.V., D.B. Spalding.A
calculation procedure for heat, mass and
momentum transfer in three-
dimensional parabolic flows.
International Journal of Heat and Mass
Transfer, 15 (1972), pp. 1787–1806

[22] Nielsen Peter V, Francis Allard,
Hazim B Awbi, Lars Davidson, and
Alois Schälin (2007) Computational
Fluid Dynamics in Ventilation Design
REHVA Guidebook No 10. International
Journal of Ventilation: 1 December
2007, Vol. 6, No. 3, pp. 291-293.

[23] Pohlhausen E., Der Wärme
austausch zwischen festen Körpern und
Flüssigkeiten mit kleine Reibung und
kleiner Wärmeleitung, Z. angew. Math.
Mech. 1 (1921).

[24] Box, G.E.P., Hunter, W.G., Hunter,
J.S. Estadística para Investigadores.
Introducción al Diseño de
Experimentos, Análisis de Datos y
Construcción de Modelos. Editorial
Reverte, 1988.

Chapter 4

Computational Approaches in Industrial Centrifugal Pumps

Atiq Ur Rehman, Akshoy Ranjan Paul, Anuj Jain
and Suvanjan Bhattacharyya

Abstract

The growing energy demand is expected to be met with increased oil and gas production. Hence, there is a need to design high-performance industrial centrifugal pumps. Recent improvements in CFD are considered as a valuable research tool to investigate the flow inside the pump and its influence on the performance of the centrifugal pump. The scope of the chapter is to emphasize the use of CFD and theoretical analysis for design and to show the prospect of improving the efficiency of a centrifugal pump. The chapter discusses the computational approaches to the CAD modeling and CFD simulation of the industrial centrifugal pumps, and the strategies and methodologies adopted. The chapter would be relevant and useful to both the pump designers, manufacturers, and industrial users.

Keywords: centrifugal pump, CFD, geometric modeling, computational modeling, NPSH, cavitation

1. Introduction

Centrifugal pump has wide application areas as it is being used in the sectors like agriculture, oil and natural gas, public water supply, sanitation, domestic, household utilities, petroleum refining, petrochemicals, mining, etc. That is why the centrifugal pump has a significant effect on the nation's economy. The present chapter describes the steps in detail that include the design of industrial centrifugal pumps, construction of geometrical models of various components of centrifugal pumps, grid generation, governing equations, boundary conditions, discretization, CFD solver settings, and the importance of CFD in improving pump design and performance.

Despite wide applications, the design of the pump and its performance prediction using a conventional approach involves trial and error which is a time-consuming task. CFD is a robust tool for prediction of performance of the centrifugal pumps that can be used to analyze and design the pump for performance and cavitation prediction [1], and condition monitoring [2] with a better understanding of flow physics. CFD analysis of any problem of engineering interest consists of three major steps, namely, pre-processing, simulation, and post-processing.

2. Pre-processing for centrifugal pump

Pre-processing consists to define the domain of interest (i.e., creation of geometry), then dividing it into a finite number of elements known as computational grid or mesh generation of the flow domain, and formulation of the problem.

2.1 Geometrical modeling of a centrifugal pump

Modeling is a process to create the complex engineering model of an industrial machine or device that can be shaped into the desired model by assembling the different parts (called solid volumes) and thus the final solid model is a virtual replica of the actual product that can be functioned (for example, rotated) like a real product. Therefore, such a complex geometrical machine or device can be modeled (constructed) by any computer-aided design (CAD) software, such as CATIA, Solid works, ProE, AutoCAD, and also with the help of non-CAD software, like BladeGen, Ansys Design Modeler (DM), etc. Modeling of an industrial centrifugal pump is always a tedious task for engineers as it deals with complex and intricate shapes associated with the impeller and volute geometries [3, 4].

A two-dimensional geometry of a single-suction, single-stage industrial centrifugal pump having a double-volute casing and impeller with three blades is generated as per the drawings made available with the help of CAD software- CATIA. The two-dimensional geometry of the impeller and double volute of an industrial pump is then converted into three-dimensional geometry for further processing. The major dimensions of the centrifugal pump are furnished in **Figure 1** shows the impeller and volute casing as separate entities, assembled and cut section of the centrifugal pump.

The three-dimensional pump geometry constructed in CATIA is imported to Ansys Design Modeler (DM) through Solid works using appropriate format to avoid

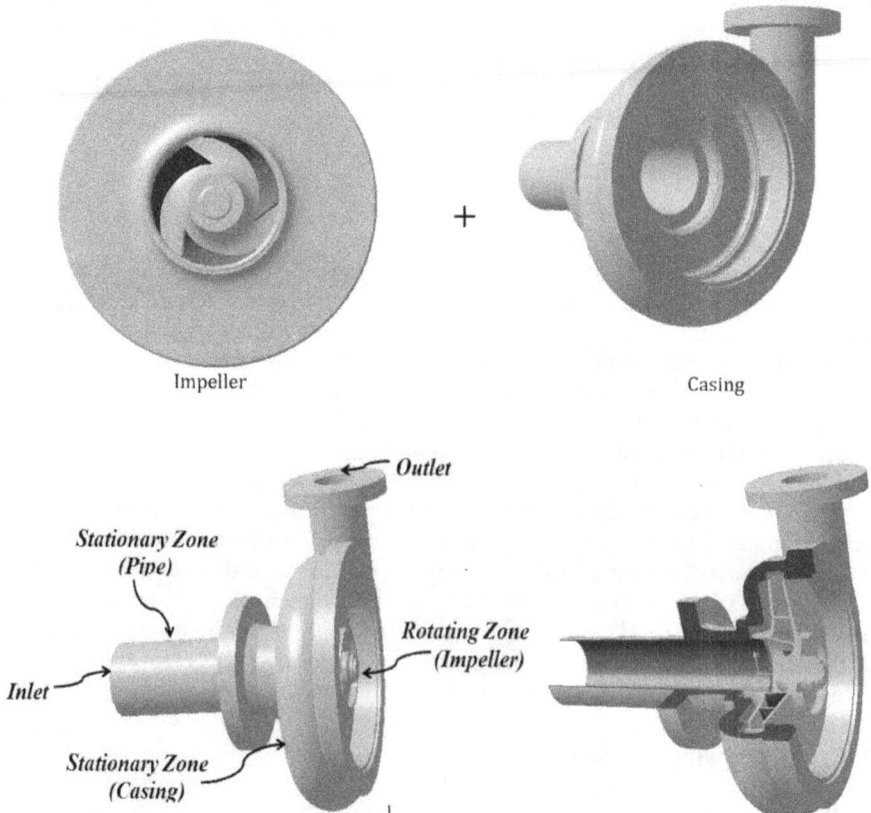

Impeller   +   Casing

Figure 1.
Impeller and casing assembly with cut-section of the centrifugal pump.

loss of surface in the process and to ensure the accurate geometry. Next, the flow domain is extracted from the imported geometry after using capping option available in Ansys DM. The extracted flow domain requires cleanup before the meshing. A flow chart explaining the steps involved in the complete process is shown in **Figure 2**.

Wireframe of the flow domain after cleanup is shown in **Figure 3** while parts of the final flow domain are shown in **Figure 4** which is used for further processing.

A pump impeller consists of an array of blades (or vanes) arranged in a certain fashion and covered with the shrouds on either side. In order to study the effect of blade geometry on the performance of the centrifugal pump, pump impellers are constructed with six different blade outlet angles. The baseline impeller consists of three backward curved-twisted blades with a 16° outlet angle. The blade thickness is varied from 6 mm at the leading edge to 3 mm at the trailing edge with a wrap angle of 278°. The construction of the impeller blade with different blade outlet angle in CATIA software is a daunting and time-consuming task. Hence, many researchers [5–8] relied on BladeGen tool available in the Ansys Blade Modular software in modeling the blade geometry of turbomachines. A two-dimensional meridional view of the impeller blade is shown in **Figure 5**, which is referred to as baseline profile. The top layer of the profile is termed as the shroud layer, while the bottom layer is known

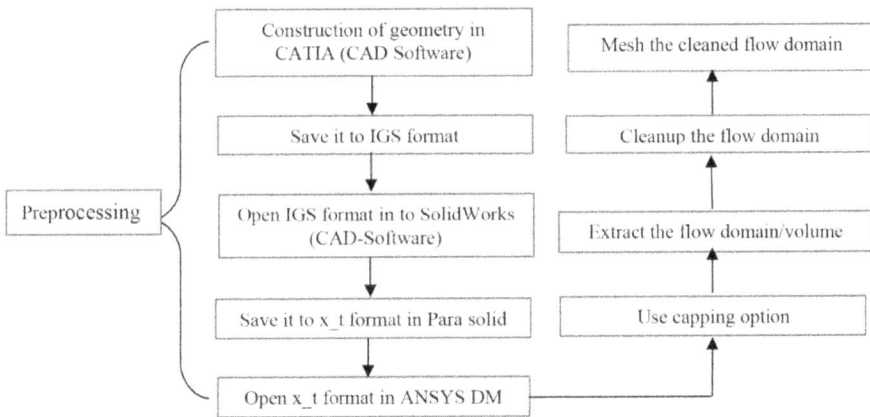

Figure 2.
Steps in the import of the geometry and extraction of the flow domain.

Figure 3.
Wireframe of the flow domain of the centrifugal pump after cleanup.

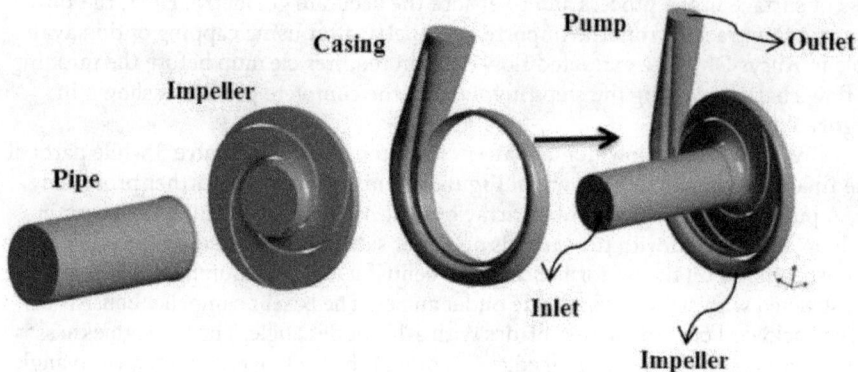

Figure 4.
Flow domain of the centrifugal pump.

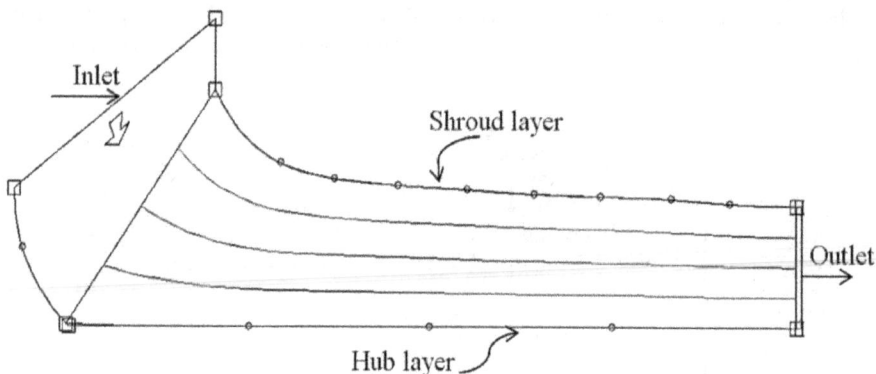

Figure 5.
Two-dimensional master profile of the impeller blade.

as the hub layer. The liquid flows around the blade profile radially and discharges at the outlet. Three-dimensional profile of a single blade of impeller is shown in **Figure 6a**, while the geometry of three blades of the impeller is shown in **Figure 6b**. The geometry of the blades thus created is imported from BladeGen to Ansys-Workbench. The extracted flow domain of a three bladed impeller is shown in **Figure 7a**.

2.2 Generation of computational grid

Grid generation (or meshing) is the next step required to perform after extraction of flow domain. It is the process of dividing the given flow domain into a number of small domains or parts called control volumes. It is seen from the literature survey that most of the researchers [8–19] used unstructured tetrahedral elements for grid generation of the centrifugal pumps. Some of the researchers [20, 21] used both structured and unstructured element, while others [22] used combined unstructured hexahedral and tetrahedral elements for centrifugal pump. Zhang et al. [23] used tetrahedral grids for meshing double-volute casing while impeller was meshed with hexahedral grids.

Due to complex profile of the impeller and the volute casing, unstructured tetrahedral grid is used in the present study for all flow domains of the centrifugal

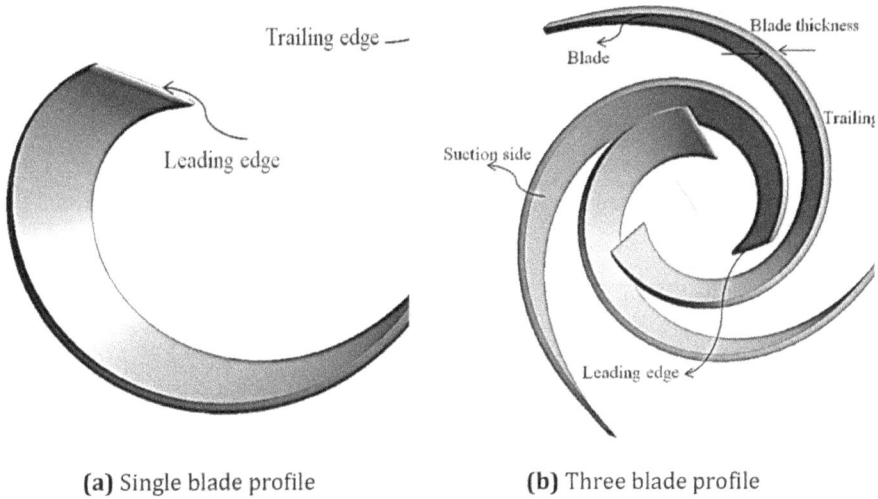

(a) Single blade profile

(b) Three blade profile

Figure 6.
Generation of blade profile of the centrifugal pump.

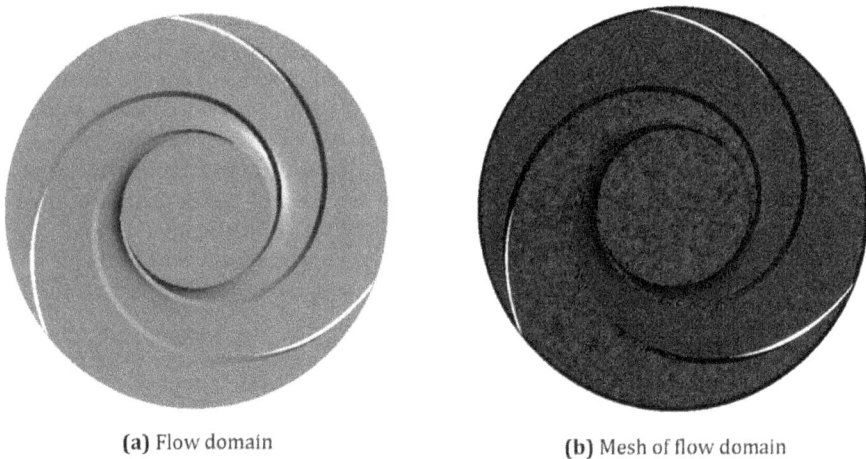

(a) Flow domain

(b) Mesh of flow domain

Figure 7.
Flow domain of and meshing of three bladed impeller.

pump using grid generation software available in Ansys-Workbench. Patch independent method has been used, which performs meshing from surface to volume,
to maintain the quality of the grid to a great extent. The accuracy of the computational results depends upon the grid density and quality. Therefore, a high-density
and high-quality grid is desirable to capture the complex flow phenomena, like flow
separation and recirculation in the diffusing passages accurately [24]. However,
high grid density increases computational time and memory. Therefore, a tradeoff
between the computational cost and the grid density is required.

Grid independence test is performed to ensure the independency of solution
with the grid size. This test is performed by refining the grid using patch independent algorithm. The number of elements are gradually increased from 1.6 million to
5.2 million in the present study to carry out the grid independence test. The results
of the grid independence test are presented in **Table 1**. The results show that no
significant change in the computed value of the head is observed from 4.2 to 5.2

Grid size	Average element size (mm)	CPU time (min.)	No of iteration	Head (m)
1,628,749	5.28	52	245	215.72
2,162,341	4.9	67	347	214.91
2,833,632	4.39	84	458	213.62
3,682,491	4.02	135	667	213.08
4,164,202	**3.86**	**181**	**804**	**210.55**
4,725,137	3.7	224	976	210.46
5,152,315	3.56	258	1014	210.56

Table 1.
Grid independence study.

million grid size. Therefore, in order to save computational time, the grid with around 4.16 million elements is chosen for further simulation. The corresponding average element size is 3.86 mm and the convergence time is 181 minutes.

Apart from size of element, the quality of a computational grid is also determined by the shape of the individual cells in the computational flow domain. The skewness of the element describes the shape of the element. CFD results are greatly affected by skewness and are therefore considered as one of the main measures for grid quality. Ansys-CFX provides a general guideline to determine the grid quality based on the value of skewness as reproduced in **Table 2** for ready reference.

It is not always easy to reduce the skewness of all the elements in the acceptable limit as indicated by **Table 1** in the complex flow domain as in the case of centrifugal pump. The literature [17, 23, 25, 26] also reports maximum skewness of centrifugal pumps' grid between 0.82 to 0.95, indicating poor to bad quality of the grid as per **Table 1**. However, it is not only the high skewness of the elements but their number and location are also important. It is found that if the highly skewed elements are few and are away from the region of interest, they do not bring a significant effect on the computational results. Histogram in **Figure 8** shows the

Value of Skewness	0.0–0.25	0.25–0.50	0.50–0.75	0.75–0.90	0.90–0.99	1.00
Grid quality	Excellent	Good	Acceptable	Poor	Bad	Degenerate

Table 2.
Range of skewness and cell quality.

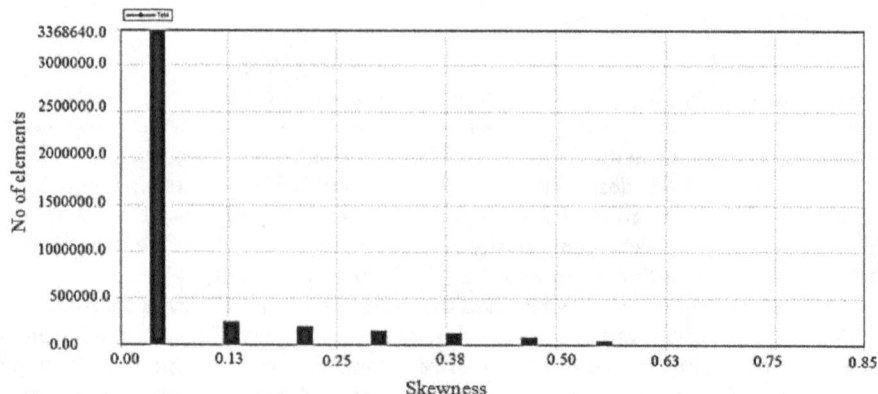

Figure 8.
Histogram of number of elements versus skewness.

frequency distribution of the skewness of the elements corresponding to 4.2 million elements in the present study. The histogram shows that insignificant number of elements are present above 0.5 skewness and almost negligible above 0.75 skewness.

The maximum skewness reported in the present study is 0.852 (affecting only 29 elements) and their locations are shown in **Figure 9**. It is observed in the present study that their presence does not influence the computational results. The final grid corresponding to 4.2 million elements is thus selected for further simulation as shown in **Figures 10** and **11**.

2.3 Governing equations

Various governing equations used in the present study to simulate the performance of the centrifugal pump are discussed in this section.

2.3.1 Continuity equation

$$\frac{\partial \rho}{\partial t} + \nabla \bullet (\rho \boldsymbol{u}) = 0 \qquad (1)$$

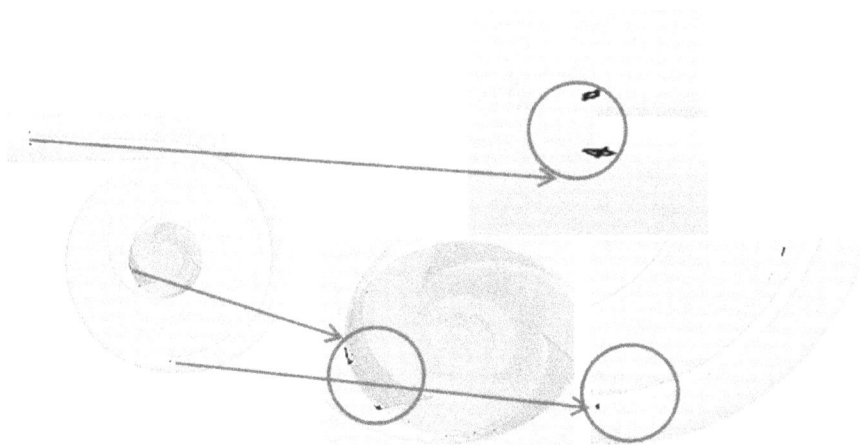

Figure 9.
Most skewed elements in the computational domain of the centrifugal pump.

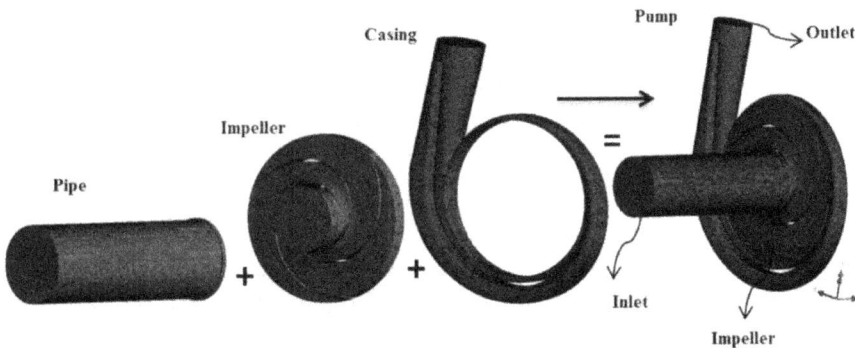

Figure 10.
Grid in the different flow domains of the centrifugal pump.

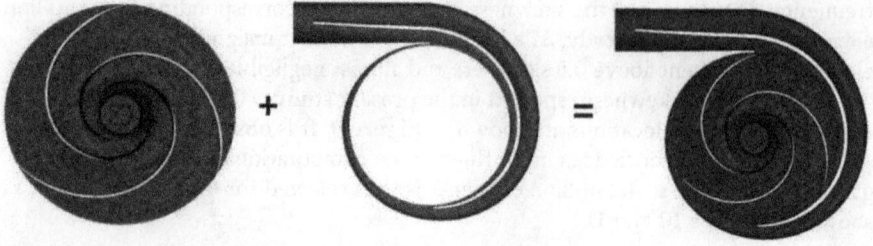

Figure 11.
(a) Grid in the impeller and casing of centrifugal pump.

2.3.2 Momentum equations

$$\frac{\partial(\rho u)}{\partial t} + \nabla \bullet (\rho u \otimes u) = -\nabla p + \nabla \bullet \tau + S_M \tag{2}$$

where τ is the stress tensor related to the strain rate by.

$$\tau = \mu_{eff}\left(\nabla u + (\nabla u)^T - \frac{2}{3}\delta\ \nabla \bullet u\right) \tag{3}$$

where S_M is the sum of body forces. For the study of flow in a rotating frame of reference having constant angular velocity ω, additional sources of momentum are required to include the effect of the Coriolis force and that of the centrifugal force:

$$S_{M,rot} = S_{Cor} + S_{cfg} \tag{4}$$

where,

$$S_{Cor} = -2\rho\omega \times u \tag{5}$$
$$S_{cfg} = -\rho\omega \times (\omega \times r) \tag{6}$$

where r is the location vector and u is the relative frame velocity.

μ_{eff} is the effective viscosity accounting the turbulence through turbulence model described in next section.

For the simulation of steady case, time derivative term will be zero in Eqs. (1) and (2).

2.3.3 Turbulence model

Various turbulence models are available to capture the turbulence behavior of the flow. Thus, the selection of appropriate turbulence model for CFD analysis of the centrifugal pump is a difficult task. The literature review reveals that most researchers used two-equation turbulence models in turbo machinery application as they offer a good compromise between computational accuracy and computational effort. A number of researchers [2, 5, 6, 9, 12, 20–24, 27, 28] used the standard k-ε turbulence model. Shahin et al. [27] and Jafarzadeh et al. [20] have used renormalized group (RNG) k-ε turbulence model. Siddique et al. [19], Deshmukh et al. [8], Paul et al. [28] and Bhattacharyya et al. [29–35] used the shear stress transport (SST) turbulence model. Alemi et al. [25] used the standard k-ε, low-Re k-ω, and SST turbulence models. Thus, it is evident that the standard k-ε turbulence model is widely used as compared to any other turbulence model for the simulation of turbomachines and other simulations involving complex geometries. Few researchers used the RNG k-ε model as an alternative to the standard k-ε model. Ansys CFX-solver theory guide suggests that the k-ω based SST model is more accurate for the prediction of the onset and the magnitude of flow separation under adverse pressure gradients by the inclusion of transport effects into the formulation of the eddy-viscosity.

Thus, four well-known two-equation turbulence models, namely standard k-ε, RNG k-ε, Wilcox k-ω, and SST turbulence models, are tested in the present study to select the one for the present study. Keeping all other conditions same, the results for these four turbulence models are obtained and compared with experimental results in **Table 3**.

It is found that predicted result by k-ω model is in more agreement with the experimental results compared to the other turbulence models. But the computation time is high as compared to others. Since standard k-ε turbulence model has taken minimum computation time and its results are also in reasonable agreement with the experimental ones, standard k-ε turbulence model is selected for further simulation in the present study. The standard k-ε model used in the present study is mentioned in the following equations. Since the standard k-ε model is based on the eddy viscosity concept,

$$\mu_{eff} = \mu + \mu_t \tag{7}$$

where μ_t is the turbulence viscosity. The k-ε model assumes that the turbulence viscosity (μ_t) is linked to the turbulence kinetic energy (k) and dissipation rate (ε) via the relation.

$$\mu_t = C_\mu \rho \frac{k^2}{\varepsilon} \tag{8}$$

$$\frac{\partial(\rho k)}{\partial t} + \frac{\partial}{\partial x_j}(\rho U_j k) = \frac{\partial}{\partial x_j}\left[\left(\mu + \frac{\mu_t}{\sigma_k}\right)\frac{\partial k}{\partial x_j}\right] + P_k - \rho\varepsilon + P_{kb} \tag{9}$$

$$\frac{\partial(\rho\varepsilon)}{\partial t} + \frac{\partial}{\partial x_j}(\rho U_j \varepsilon) = \frac{\partial}{\partial x_j}\left[\left(\mu + \frac{\mu_t}{\sigma_\varepsilon}\right)\frac{\partial\varepsilon}{\partial x_j}\right] + \frac{\varepsilon}{k}(C_{\varepsilon1}P_k - C_{\varepsilon2}\rho\varepsilon + C_{\varepsilon1}P_{\varepsilon b}) \tag{10}$$

where model constants are $C_\mu = 0.09, \ C_{\varepsilon1} = 1.44, C_{\varepsilon2} = 1.92, \sigma_k = 1.0$ and $\sigma_\varepsilon = 1.3$.

P_{kb} and $P_{\varepsilon b}$ represent the influence of the buoyancy forces associated with k and ε, respectively.

P_k is the turbulence production term due to viscous force, which is modeled for incompressible flow as.

Inlet boundary condition	Outlet boundary condition	Turbulence model	Head (m)	Output power (kW)	Input power (kW)	Efficiency (%)
Total pressure (Pa)	Mass flow rate (kg/s)	Standard k-ε	210.55	138.84	176.27	78.76
		RNG k-ε	210.49	138.80	176.15	78.79
		k-ω	203.53	137.19	174.00	78.84
		SST k-ω	204.62	134.93	172.57	78.18
	Experimental results		206.54	136.65	185.11	73.82

Table 3.
Comparison of turbulence models for Q = 242.8 m³/h at 2933 rpm.

$$P_k = \mu_t \left(\frac{\partial U_i}{\partial x_j} + \frac{\partial U_j}{\partial x_i} \right) \frac{\partial U_i}{\partial x_j} - \frac{2}{3} \frac{\partial U_k}{\partial x_k} \left(3\mu_t \frac{\partial U_k}{\partial x_k} + \rho k \right) \qquad (11)$$

Scalable wall function is used to capture near wall flow physics in k-ε turbulence model. Judicious choice of y^+ value affects the quality of computed results as well as computation time [36]. This approach limits the value of y^+ used in the logarithmic formulation as follows:

$$y^+ = max\,(y^*, h_s^+/2, 11.06) \qquad (12)$$

This approach also incorporates the effect of wall surface roughness as dimensionless sand-grain roughness, h_s^+.

2.3.4 Turbulence intensity

The turbulence intensity is defined as the ratio of root mean square of fluctuating velocity (u') to the mean flow velocity ($u_{avg.}$). For internal flow, it is found that turbulence intensity less than 1% represents low turbulence and turbulence intensity greater than 10% represents high turbulence. Thus, the allowable range of turbulence intensity is from 1–10% corresponding to very low to very high levels of turbulence in the internal flow. The turbulence intensity in the fully developed duct flow can be calculated from the following formula:

$$I = \frac{u'}{u_{avg.}} = 0.16\,(Re_{D_H})^{-1/8} \qquad (13)$$

Nominal turbulence intensity ranges from 1–5% depending up on the specific application. In absence of experimental data, the default turbulence intensity value of 3.7% is considered good estimate for nominal turbulence through a circular inlet.

It is found that many researchers [25, 37–41] used 5% turbulence intensity at the inlet boundary conditions for turbomachines, while Shahin et al. [27] used 2% turbulence intensity, Jafarzadeh et al. [20] used 1% turbulence intensity, Murmu et al. [42] and Alam et al. [43] investigated varied turbulence intensity at the inlet boundary condition.

Three standard turbulence intensity, 1%, 5%, and 10%, are tested in the present study for CFD simulation of flow through the centrifugal pump. The results are presented in **Table 4** for comparison. The results show almost negligible variation in the value of head, output power, input power, and efficiency of centrifugal pump. Hence turbulence intensity equal to 5% is finally chosen for further study.

2.3.5 Boundary conditions

While solving the Navier–Stokes equations and continuity equation, an appropriate boundary conditions need to be applied. Boundary conditions are a very

Turbulence intensity (%)	Head (m)	Output power (kW)	Input power (kW)	Efficiency (%)
1.0	210.5	138.83	176.26	78.76
5.0	210.5	138.83	176.26	78.76
10.0	210.5	138.84	176.27	78.77

Table 4.
Comparison of turbulence intensity for Q = 242.8 m³/h at 2933 rpm.

important set of the properties and conditions applied on the surfaces of computational domains. These are required to define the flow field completely. The solid and fluid regions are generally represented by domains or cell zones while internal surfaces and boundaries are represented by the face zones. The data corresponding to boundary conditions are then assigned to these face zones. Each variable equation requires the meaningful values at the boundaries at the domain in order to generate the values in entire domain. Boundary conditions used by various researchers working in computational study of centrifugal pump are summarized in **Table 5**.

After analyzing all the boundary conditions, it is found that the computational results of centrifugal pump corresponding to the boundary condition of total

Authors	Year	Inlet	Outlet
Blanco-Marigorta et al. [26]	2000	Flow rate	Static pressure.
Stickland et al. [9]	2000	Flow rate	Static pressure.
Gonzalez et al. [10]	2002	Total pressure	A variable static pressure proportional to the kinetic energy at the outlet for their unsteady pump.
Majidi [37]	2005	Mass flow rate	For all variables (with exception of pressure) a zero-gradient condition.
Gonzalez and Santolaria [13]	2006	Total pressure	A pressure drop proportional to the kinetic energy at the outlet.
Cheah et al. [14]	2007	Mass flow rate	Mass flow is being specified.
Cheah et al. [38]	2008	Total pressure	Mass flow rate.
Spence and Teixeira [44]	2009	Mass flow rate	Static pressure.
Feng et al. [45]	2010	Total pressure	Mass flow rate.
Perez et al. [16]	2010	Relative Pressure of 0 Pa	Mass flow rate.
Houlin et al. [17]	2010	Velocity	Outflow.
Li [46]	2011	Velocity	Mass flow rate
Jafarzadeh et al.[20]	2011	Volume flow rate	Pressure.
Hedi et al. [47]	2012	Total pressure of 1 atm	Mass flow rate.
Kumar et al. [22]	2013	Maas flow rate	Static pressure.
Bellary and Samad [5]	2013	Pressure	Mass flow rate.
Lei et al. [48]	2014	Velocity	Pressure.
Kim et al. [40]	2014	Pressure of 0 atm	Mass flow rate.
Li [41]	2014	Absolute velocity	Zero Static pressure.
Bellary and Samad [49]	2015	Pressure	Mass flow rate.
Alemi et al. [25]	2015	Mass flow rate	Pressure.
Siddique et al. [19]	2017	Static pressure	Mass flow rate.
Deshmukh et al. [8]	2017	Static pressure of 1.01 kPa	Mass flow rate.

Table 5.
Boundary conditions used by researchers for centrifugal pump.

pressure at inlet and mass flow rate matches most closely with the experimental measurement. The inlet total pressure is known from the measured data corresponding to different discharge. The turbulence kinetic energy and turbulence dissipation rate at the inlet totally depends on the upstream history of the flow.

The inlet pipe and casing are considered as stationary reference frames while impeller is considered as rotating reference frame. These frames are coupled through frozen-rotor interface model. There are two interfaces, one connecting pipe and impeller and other connecting impeller and casing. Walls associated with impeller are considered as rotating walls while those associated with pipe and casing are considered as stationary walls. No-slip boundary conditions are applied over the impeller blades and casing walls. A list of boundary conditions used for the CFD simulation of the pump in the study is given in **Table 6**.

2.4 CFD solver settings

The governing equations along with boundary conditions are discretized using second order upwinding scheme. The discretized equations result in a set of algebraic equations. Ansys-CFX, an element based finite volume method (EbFVM) solver with a cell vertex formulation is used for the present study to solve the algebraic equations. **Table 7** shows various solver settings used during the CFD simulation.

2.4.1 Convergence criteria

Ansys-CFX uses a multigrid (MG) accelerated incomplete lower upper (ILU) factorization technique for solving the discretized system of equations. This technique is iterative in nature. The measure of convergence of iterative solution is the residual which quantifies the error in the computational solution. In a CFD analysis, the residual quantifies the local imbalance of each conservative control volume

Parameters	Description
Inlet	Total Pressure
Outlet	Mass flow rate
Blades	Rotating wall
Pipe	Stationary reference frame
Impeller	Rotating reference frame
Casing	Stationary reference frame
Interface-1 (Between pipe outlet and impeller inlet)	Fluid–Fluid
Interface-2 (Between impeller outlet and casing inlet)	Fluid–Fluid

Table 6.
Boundary conditions used.

Solver	Coupled
Analysis method	Multiple reference frame (MRF) or Frozen rotor method
	Steady/Unsteady state analysis
Reference Pressure	1 atm

Table 7.
Solver setting options used in Ansys-CFX.

equation. Thus, after specifying all the boundary conditions, numerical schemes, convergence criteria are specified to solve the problem.

Researchers used various convergence criteria for numerical simulation of centrifugal pumps. Feng et al. [45] taken convergence criteria 10^{-3}. Majidi [37], Cheah et al. [14], Perez et al. [16], Hedi et al. [47], Li [46], Alemi et al. [25], Siddique et al. [19] used 10^{-4} residual, while Gonzalez et al. [10] used a residual of 10^{-5} for mass and momentum and 10^{-4} for k-ε turbulence model. Gonzalez et al. [9], Houlin et al. [17], Bellary and Samad [5, 28], Deshmukh et al. [8] have considered the residual at 10^{-5}.

The computational effort and computational accuracy are always competing parameters. Thus, the accuracy of the computation in terms of root mean square (RMS) of residual up to10^{-6} is calculated with the corresponding computational effort for two different turbulence models, k-ε and k-ω, at a flow rate of 242.8 m³/h and rotation speed of impeller equal to 2933 rpm. Keeping computation time in consideration, the convergence criteria for further run of flow simulation is set to 10^{-5} for residuals of mass and momentum equations.

2.4.2 Steady state flow simulation

Moving reference frame (MRF) allows the computational analysis of cases involving domains that are rotating relative to one another. Since MRF is based on the general grid interface (GGI) technology, this feature allows rotor/stator interaction in the investigation of turbomachines in Ansys-CFX. In GGI, a control surface approach is used to connect across the interface or periodic condition. This intersection algorithm allows the complete freedom to change the physical distribution and the grid topology across the interface. This permits the use of most appropriate meshing style for each component involved in the analysis. Many researchers [14, 15, 18, 20, 27, 39, 41, 45, 50] carried out computational investigation of centrifugal pump to assess its performance using MRF approach in Ansys-Fluent and Ansys-CFX.

The steady state CFD simulation is carried out using MRF approach considering the volute and pipe sections as stationary frame while impeller flow field as rotating frame. In this case, the relative position between volute casing and impeller flow field is considered fixed in time and space. The grids of pipe section, impeller, and volute that are generated separately with the help of tetrahedral elements are connected by means of a "frozen-rotor" interface. The interface treatment is fully implicit that preserves the flow field variation across the interface and does not adversely affect convergence of the overall solution.

2.4.3 Unsteady flow simulation

Transient/unsteady behavior can be caused by changing the boundary conditions of the flow or can be related to the flow where steady state condition is never reached, even when all other aspects of the flow conditions are not changing. For the unsteady simulation of the present problem, the surface fluxes at each side of the interface are first computed at the start of each time step at the current relative position. The dissimilar meshes at the pipe, impeller, and volute interfaces are connected by GGI connections to permit transient flow interaction between a stator and rotor passages across the sliding (transient rotor-stator/frame change) interface. This approach allows the accounting of all interaction effects between the components that are in the transient relative motion to each other across the interface. The interface position is updated at each time step as the relative position of the grid's changes on each side of the interface. To determine the real time information, the transient simulations require appropriate time interval at which the Ansys-CFX solver calculates the flow field.

A few researchers [10, 22, 26, 27, 37, 42] also studied the transient CFD simulation. Three different time step sizes (3°, 6°, 9°) are compared for Q = 280 m³/h at 2933 rpm in the present study to choose the appropriate time step size which provides the necessary time resolution as shown in **Table 8**. From the study, it is found that 6° impeller rotation per step is enough to reduce the maximum residuals to 10^{-5}. Thus, impeller rotation step size of 6° is chosen for further study corresponds to 60 steps/rotation and 360 steps to complete 6 full rotations of the impeller corresponding to the pump running time equal to 0.12274 s.

2.4.4 Cavitation simulation using Rayleigh–Plesset equation

In the present study, the steady state numerical simulation with cavitation is carried out with a multiple frames of reference (MRF) approach using the Rayleigh Plesset equation to investigate the cavitation in the centrifugal pump. The Rayleigh–Plesset equation provides the basis for the rate equation controlling vapor generation and condensation. The Rayleigh–Plesset equation that describes the growth of a gas bubble in a liquid is derived from a mechanical balance, assuming no thermal barriers to bubble growth. The equation is given as:

$$R_B \frac{d^2 R_B}{dt^2} + \frac{3}{2}\left(\frac{dR_B}{dt}\right)^2 + \frac{2\sigma}{\rho_f R_B} = \frac{p_v - p}{\rho_f} \tag{14}$$

where R_B is the bubble radius, p_v is the pressure in the bubble (assumed to be equal to the vapor pressure at the temperature of the liquid), p is the liquid pressure surrounding the bubble, ρ_f is the density of liquid, and σ is the coefficient of surface tension between the liquid and vapor. Neglecting the surface tension and the second order terms (which is acceptable for low oscillation frequencies), the equation reduces to:

$$\frac{dR_B}{dt} = \sqrt{\frac{2}{3}\frac{p_v - p}{\rho_f}} \tag{15}$$

The rate of change of bubble volume is:

$$\frac{dV_B}{dt} = \frac{d}{dt}\left(\frac{4}{3}\pi R_B^3\right) = 4\pi R_B^2\sqrt{\frac{2}{3}\frac{p_v - p}{\rho_f}} \tag{16}$$

thus, the rate of change of bubble mass is:

$$\frac{dm_B}{dt} = \rho_g\frac{dV_B}{dt} = 4\pi R_B^2\rho_g\sqrt{\frac{2}{3}\frac{p_v - p}{\rho_f}} \tag{17}$$

Impeller rotation/step	Total no of time steps/6 revolutions	Time taken/ revolution (ms)	Time per step size (ms)	Total time taken for 6 revolution (ms)
9°	240	20.46	0.51	122.74
6°	360	20.46	0.34	122.74
3°	720	20.46	0.17	122.74

Table 8.
Comparison of impellerrelative positions at Q = 280 m³/h at 2933 rpm.

The volume fraction r_g may be expressed as:

$$r_g = V_B N_B = \frac{4}{3}\pi R_B^3 N_B \tag{18}$$

where N_B is number of bubbles per unit volume,
and the total inter-phase mass transfer rate per unit volume is:

$$\dot{m}_{fg} = N_B \frac{dm_B}{dt} = \frac{3r_g\rho_g}{R_B}\sqrt{\frac{2}{3}\frac{p_v - p}{\rho_f}} \tag{19}$$

This expression has been derived assuming bubble growth due to vaporization. The expression can be generalized to include condensation as follows:

$$\dot{m}_{fg} = F\frac{3r_g\rho_g}{R_B}\sqrt{\frac{2}{3}\frac{p_v - p}{\rho_f}}\, \text{sgn}\,(p_v - p) \tag{20}$$

where F is an empirical factor that is designed to account for rates of condensation and vaporization. Vaporization is usually much faster than condensation. Thus, F has different values for vaporization and condensation. Despite the fact that Eq. (20) has been generalized for vaporization and condensation, it requires further modification in the case of vaporization. Vaporization is initiated most commonly at the nucleation sites for non-condensable gases. Thus, the bubble radius R_B will be replaced by the nucleation site radius R_{nuc} for the purpose of modeling of vaporization. The nucleation site density decreases as the vapor volume fraction increases as there is less liquid available. For the case of vaporization, r_g in Eq. (20) is replaced by $r_{nuc}(1 - r_g)$ to give mass transfer rate as:

$$\dot{m}_{fg} = F\frac{3r_{nuc}(1 - r_g)\rho_g}{R_{nuc}}\sqrt{\frac{2}{3}\frac{|p_v - p|}{\rho_f}}\, \text{sgn}\,(p_v - p) \tag{21}$$

where r_{nuc} is the nucleation site volume fraction. Eq. (21) is maintained in the case of condensation.

To obtain an inter-phase mass transfer rate, the values of bubble concentration and radius are required. The Rayleigh–Plesset cavitation model uses the following default values for the model parameters as implemented in Ansys-CFX:

$$R_{nuc} = 1\mu m; \quad r_{nuc} = 5E - 4; \quad F_{vap} = 50; \quad F_{cond} = 0.01$$

2.4.5 Acoustics simulation

Noise in the pump is produced primarily by the unsteady pressure field. To predict mid- to far-field noise, the methods based on Lighthill's acoustic analogy offer sustainable alternative to the direct method. The acoustic analogy decouples the generation of sound from its propagation. Thus, enabling the separation of the flow solution process from the acoustics analysis. Therefore, in this approach, the near-field flow is obtained from the appropriate flow governing equations such as unsteady RANS equations. Then the sound is predicted with the help of analytically derived integral solutions of wave equations using determined flow.

2.4.6 The Ffowcs-Williams and Hawkings Equation for Acoustic Study

Ansys-Fluent offers an acoustic study method based on the Ffowcs-Williams and Hawkings (FW-H) formulation. The Ffowcs-Williams and Hawkings (FW-H) equation is essentially an inhomogeneous wave equation that is derived by manipulating the Navier–Stokes equations and the continuity equation. The FW-H formulation uses the most general form of Light hill's acoustic analogy neglecting the interaction between fluid and solid. This formulation predicts sound generated by equivalent acoustic sources. ANSYS-Fluent uses a time domain integral formulation to compute time histories of sound pressure, or acoustic signals, at prescribed receiver locations. The FW-H equation can be written as:

$$\frac{1}{a_0^2}\frac{\partial^2 p'}{\partial t^2} - \nabla^2 p' = \frac{\partial^2}{\partial_i \partial_j}\{T_{ij}H(f)\} - \frac{\partial}{\partial x_i}\left\{\left[p_{ij}n_j + \rho u_i(u_n - v_n)\right]\delta(f)\right\}$$
$$+ \frac{\partial}{\partial t}\{[\rho_0 v_n + \rho(u_n - v_n)]\delta(f)\}$$

(22)

where
u_i = fluid velocity component in the X_i direction
u_n = fluid velocity component normal to the surface $f = 0$
v_i = surface velocity components in the X_i direction
v_n = surface velocity component normal to the surface
$\delta(f)$ = Dirac delta function
$H(f)$ = Heaviside function
$p' = p - p_0$ is the sound pressure at the far field.
In order to facilitate the use of generalized function theory and the free space Green function to obtain the solution, a mathematical surface represented by $f = 0$ is introduced to 'embed' the exterior flow problem ($f > 0$) in an unbounded space. The surface ($f = 0$) corresponds to the source (emission) surface and can be made to coincide with a body (impermeable) surface or off the body permeable surface.
n_j is the unit normal vector pointing toward the exterior region ($f > 0$),
a_0 is the far-field speed of sound,
T_{ij} is the Lighthill stress tensor, defined as

$$T_{ij} = \rho u_i u_j + P_{ij} - a_0^2(\rho - \rho_0)\delta_{ij}$$

(23)

For a Stokesi an fluid, the compressive stress tensor, P_{ij}, is given by

$$P_{ij} = p\delta_{ij} - \mu\left[\frac{\partial u_i}{\partial x_j} + \frac{\partial u_j}{\partial x_i} - \frac{2}{3}\frac{\partial u_k}{\partial x_k}\delta_{ij}\right]$$

(24)

The subscript 0 denotes the free-stream properties.

2.4.7 Proudman's formula

Proudman [51] derived a formula for acoustic power generation by isotropic turbulence without mean flow using Light hill's acoustic analogy. For a given turbulence field, the Proudman's formula yields an approximate measure of the local contribution to total acoustic power per unit volume. Proudman's original derivation neglected the retarded time difference. Later, the formula was re-derived by accounting for the retarded time difference. Both derivations provide acoustic power due to the unit volume of isotropic turbulence in (W/m^3) as

$$P_A = \alpha \rho_0 \left(\frac{u^3}{\ell}\right)\frac{u^5}{a_0^5}$$

(25)

where u is the turbulence velocity,

$\ell_=$ turbulence length scale, $a_{0=}$ speed of sound, α = model constant.
Eq. (25) can be rewritten in terms of k and ε as.

$$P_A = \alpha_\varepsilon \rho_0 \varepsilon M_t^5 \tag{26}$$

where

$$M_t = \frac{\sqrt{2k}}{a_0} \tag{27}$$

α_ε is the rescaled constant, set to 0.1

In the present study, the FW-H model is used to measure acoustics in terms of the level of noise radiation induced by the inner flow field in the centrifugal pump encompassing the impellers and volute. In FW-H acoustic model, it is essentially required that all the receivers are located far away (at least at a distance of 1 m) from the primary sources of sound. Thus, for monitoring, 24 receivers, indicated by P1-P24, are arranged in circular surface at an angular gap of 15° is shown in **Figure 12**.

Receiver P7 is placed in line with the volute tongue and receiver P19 is placed in line with the volute splitter. To extract the relevant acoustic spectra, a Fourier

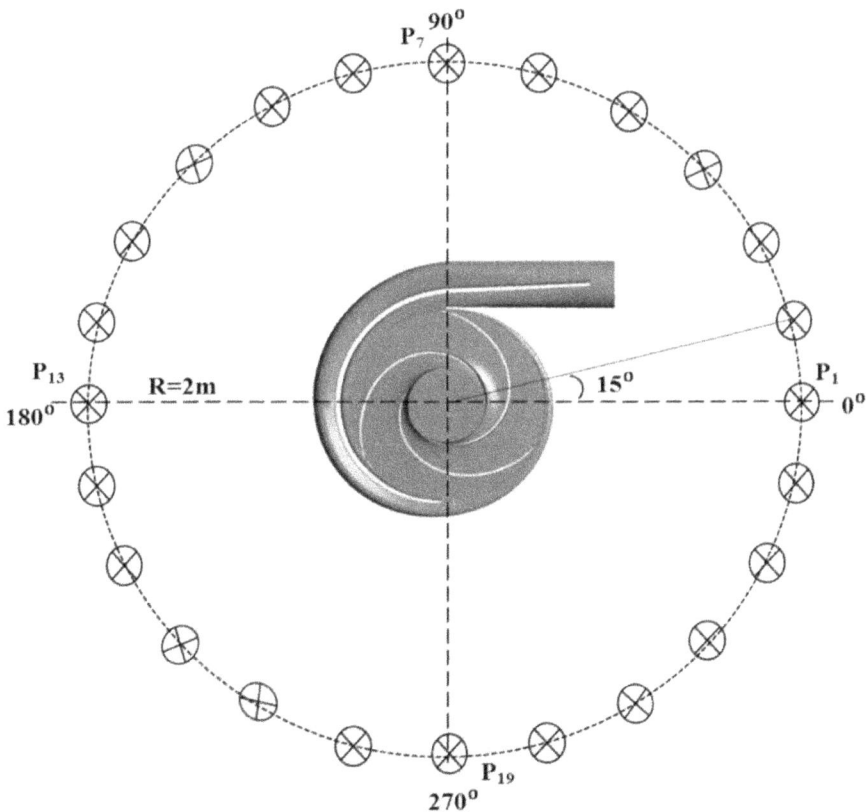

Figure 12.
Arrangement of sound receivers outside the centrifugal pump.

transformation is applied connecting the acoustic signals found at these receivers. The sound pressure level (SPL) in decibel (dB) is calculated by:

$$SPL = 20 \log_{10} \left(\frac{P_e}{P_{\mathrm{Ref}}} \right) \tag{28}$$

where P_{Ref} is the reference sound pressure = 2×10^{-5} Pa for air, P_e is the effective acoustic pressure defined by:

$$P_e = \sqrt{\frac{1}{T} \int_0^T p r^2 dt} \tag{29}$$

It is necessary to derive a temporal intensity profile to obtain the intensity of noise radiation, involving a superposition of acoustic pressures at each Fourier frequency. In this regard, the total sound pressure level(TSPL) is introduced and expressed as:

$$TSPL = 10 \lg \sum_{i=1}^{n} 10^{SPL_i / 10} \tag{30}$$

Detailed analysis of cavitation detection in a centrifugal pump impeller using acoustic generation can be found in Jaiswal et al. [52].

3. Conclusions

Intensive computation coupled with geometric modeling helps the engineers and industrial users of the centrifugal pumps to investigate its performance characteristics [53] and cavitation prediction while dealing with multiple working fluids [54] and geometric modification of blades and its number [55, 56]. The present chapter discussed the computational strategies that encompasses the CAD modeling, computational grid generation, steps involved in the CFD solvers, and the various issues associated with it.

Further studies in this field may include computational analysis on a double-suction, double-volute centrifugal pump impeller [57] to investigate the effects of double-suction on its hydraulic performance and cavitation characteristics. Besides, optimization of impeller blades using surrogate modeling [58] can also lead to design of an improved centrifugal pump impeller.

Author details

Atiq Ur Rehman[1], Akshoy Ranjan Paul[1]*, Anuj Jain[1] and Suvanjan Bhattacharyya[2]*

1 Department of Applied Mechanics, Motilal Nehru National Institute of Technology Allahabad, Pryagraj, Uttar Pradesh, India

2 Department of Mechanical Engineering, Birla Institute of Technology and Science, Pilani, Rajasthan, India

*Address all correspondence to: arpaul@mnnit.ac.in and suvanjan.bhattacharyya@pilani.bits-pilani.ac.in

IntechOpen

References

[1] Rehman AU, Vinoth Kumar K, Jain A, Paul AR. Performance and cavitation prediction in an industrial centrifugal pump for different working fluids. In: Proceedings of First International Conference on Thermal and Fluid Engineering (ICTF-2017), 3-5 July 2017, Phuket (Thailand). 2017. pp. 36-42

[2] Rehman AU, Shinde S, Singh VK, Paul AR, Jain A, Mishra R. CFD based condition monitoring of centrifugal pumps. In: Proceedings of COMADEM 2013, 11–13 June 2013, Helsinki, Finland. 2013

[3] Rehman AU, Shinde S, Paul AR, Jain A. Effects of discharge on the performance of an industrial centrifugal pump. In: Proceedings of International Conference on Computing Techniques and Mechanical Engineering (ICCTME-2015), Oct. 1–3, 2015, Phuket (Thailand). 2015

[4] Rehman AU, Paul AR, Jain A. Effect of impeller volute interaction on performance of the centrifugal pump: A CFD approach. In: Proc. of Indian Conference on Applied Mechanics (INCAM-2017) at MNNIT Allahabad, July 5-7, 2017. 2017

[5] Bellary SAI, Samad A. Exit blade angle and roughness effect on centrifugal pump performance. In: ASME 2013 Gas Turbine India Conference. 2013

[6] Bellary, S.A.I. and Samad, A., 2015-a. Pumping crude oil by centrifugal impeller having different blade angles and surface roughness. Journal of Petroleum Exploration and Production Technology, 6l, pp.117-127.

[7] Yun JE, Kim JH. Effect of surface roughness on performance analysis of centrifugal pump for wastewater transport. Transactions of the Korean Society of Mechanical Engineers B. 2014;**38**(2):147-153

[8] Deshmukh D, Siddique MH, Samad A. Surface roughness effect on performance of an electric submersible pump. In: ASME 2017 Gas Turbine India Conference. 2017

[9] Stickland MT, Scanlon TJ, Blanco-Marigorta E, Fernández-Francos J, González-Pérez J, Santolaria-Morros C. Numerical flow simulation in a centrifugal pump with impeller-volute interaction. In: Proceedings of the ASME, Fluids Engineering Division Summer Meeting. 2000

[10] Gonzalez J, Fernández J, Blanco E, Santolaria C. Numerical simulation of the dynamic effects due to impeller-volute interaction in a centrifugal pump. Journal of Fluids Engineering. 2002;**124**(2):348-355

[11] Zhou W, Zhao Z, Lee TS, Winoto SH. Investigation of flow through centrifugal pump impellers using computational fluid dynamics. International Journal of Rotating Machinery. 2003;**9**(1):49-61

[12] Gonzalez, J., Parrondo, J., Santolaria, C. and Blanco, E., 2006-a. Steady and unsteady radial forces for a centrifugal pump with impeller to tongue gap variation. Journal of Fluids Engineering, 128(3), pp.454-462.

[13] Gonzalez, J. and Santolaria, C., 2006-b. Unsteady flow structure and global variables in a centrifugal pump. Journal of Fluids Engineering, 128(5), pp.937-946.

[14] Cheah KW, Lee TS, Winoto SH, Zhao ZM. Numerical flow simulation in a centrifugal pump at design and off-design conditions. International Journal of Rotating Machinery. 2007;**83641**:1-8

[15] Bacharoudis EC, Filios AE, Mentzos MD, Margaris DP. Parametric study of a centrifugal pump impeller by varying the outlet blade angle. Open Mechanical Engineering Journal. 2008;**2**(5):75-83

[16] Perez J, Chiva S, Segala W, Morales R, Negrao C, Julia E, et al. Performance analysis of flow in a impeller-diffuser centrifugal pumps using CFD: Simulation and experimental data comparisons. In: V European Conference on Computational Fluid Dynamics EECMV, ECCOMAS CFD. 2010

[17] Houlin L, Yong W, Shouqi Y, Minggao T, Kai W. Effects of blade Numberon characteristics of centrifugal pumps. Chinese Journal of Mechanical Engineering-English Edition. 2010;**23**(6):742-747

[18] Cui B, Wang C, Zhu Z, Jin Y. Influence of blade outlet angle on performance of low-specific-speed centrifugal pump. Journal of Thermal Science. 2013;**22**(2):117-122

[19] Siddique MH, Bellary SAI, Samad A, Kim JH, Choi YS. Experimental and numerical investigation of the performance of a centrifugal pump when pumping water and light crude oil. Arabian Journal for Science and Engineering. 2017;**42**(11): 4605-4615

[20] Jafarzadeh B, Hajari A, Alishahi MM, Akbari MH. The flow simulation of a low-specific-speed high-speed centrifugal pump. Applied Mathematical Modelling. 2011;**35**(1): 242-249

[21] Hedi ML, Hatema K, Ridha Z. Numerical flow simulation in a centrifugal pump. International Renewable Energy Congress. 2010; **2010**:300-304

[22] Kumar S, Mohapatra SK, Gandhi BK. Investigation on centrifugal slurry pump performance with variation of operating speed. International Journal of Mechanical and Materials Engineering. 2013;**8**(1):40-47

[23] Zhang, Y.L., Zhu, Z.C., Dou, H.S., Cui, B.L., Li, Y. and Zhou, Z.Z., 2017-a. Numerical investigation of transient flow in a prototype centrifugal pump during Startup period. International Journal of Turbo & Jet-Engines, 34(2), pp.167-176.

[24] Sagar D. Akshoy Ranjan Paul, Anuj Jain, "computational fluid dynamics investigation of turbulent separated flows in axisymmetric diffusers", international journal of engineering. Science and Technology (IJEST). 2011;**3**(2):104-109

[25] Alemi H, Nourbakhsh SA, Raisee M, Najafi AF. Effects of volute curvature on performance of a low specific-speed centrifugal pump at design and off-design conditions. Journal of Turbomachinery. 2015;**137**(4):041009-041010

[26] Blanco-Marigorta E, Fernandez-Francos J, Parrondo-Gayo J, Santolaria-Morros C. Numerical simulation of centrifugal pumps. In: ASME Paper. 2000

[27] Shahin I, Abdelganny M, Abdellatif OE, Abdrabbo MF. Performance and unsteady flow field prediction of a centrifugal pump with CFD tools. In: Proceedings of ICFD 10: Tenth International Congress of Fluid Dynamics. 2010

[28] Paul AR, Jain A, Alam F. Drag reduction of a passenger car using flow control techniques. International Journal of Automotive Technology. 2019;**20**(2): 397-410. DOI: 10.1007/s12239-019-0039-2

[29] Bhattacharyya S, Chattopadhyay H, Benim AC. Simulation of heat transfer enhancement in tube flow with twisted

tape insert. Progress in Computational Fluid Dynamics. 2017;**17**(3)

[30] Bhattacharyya S, Dey K, Hore R, Banerjee A, Paul AR. Computational study on thermal energy around diamond shaped cylinder at varying inlet turbulent intensity. Energy Procedia. 2019;**160**(2018): 285-292

[31] Bhattacharyya S, Khan AI, Maity DK, Pradhan S, Bera A. Hydrodynamics and heat transfer of turbulent flow around a rhombus cylinder. Chemical Engineering Transactions. 2017;**62** (1987):373-378

[32] Bhattacharyya S, Benim AC, Chattopadhyay H, Banerjee A. Experimental and numerical analysis of forced convection in a twisted tube. Thermal Science. 2019;**23**

[33] Bhattacharyya S, Das S, Sarkar A, Guin A, Mullick A. Numerical simulation of flow and heat transfer around hexagonal cylinder. International Journal of Heat and Technology. 2017;**35**(2):360-363

[34] Bhattacharyya S, Chattopadhyay H, Biswas R, Ewim DRE, Huan Z. Influence of inlet turbulence intensity on transport phenomenon of modified diamond cylinder: A numerical study. Arabian Journal for Science and Engineering. 2020;**45**(2): 1051-1058

[35] Bhattacharyya S, Chattopadhyay H, Bandyopadhyay S. Numerical study on heat transfer enhancement through a circular duct fitted with Centre-trimmed twisted tape. International Journal of Heat and Technology. 2016; **34**(3):401-406

[36] Srivastav VK, Paul AR, Jain A. Capturing the wall turbulence in CFD simulation of human respiratory tract. Mathematics and Computers in Simulation. s;**160**:23-38

[37] Majidi K. Numerical study of unsteady flow in a centrifugal pump. Journal of Turbomachinery, Transactions of the ASME. 2005;**127**: 363-371

[38] Cheah KW, Lee TS, Winoto SH. Unsteady fluid flow study in a centrifugal pump by CFD method. In: 7th ASEAN ANSYS Conference Biopolis, Singapore 30th and 31st October. 2008

[39] Mentzos MD, Vouros AP, Margaris DP, Filios AE, Kaldellis JK. The effect of the blade geometry in the velocity profile in a radial pump impeller. In: 5th International Conference on Experiments/ Process/ System Modeling/ Simulation/ Optimization. 2013. pp. 1-7

[40] Kim S, Lee KY, Kim JH, Choi YS. Numerical study on the improvement of suction performance and hydraulic efficiency for a mixed-flow pump impeller. In: Mathematical Problems in Engineering. 2014.A

[41] Li WG. Mechanism for onset of sudden-rising head effect in centrifugal pump when handling viscous oils. Journal of Fluids Engineering. 2014;**136** (7):074501-074510

[42] Murmu SC, Bhattacharyya S, Chattopadhyay H, Biswas R. Analysis of heat transfer around bluff bodies with variable inlet turbulent intensity: A numerical simulation. International Communications in Heat and Mass Transfer. 2020;**117**:104779

[43] Alam MW, Bhattacharyya S, Souayeh B, Dey K, Hammami F, Rahimi-Gorji M, et al. CPU heat sink cooling by triangular shape micro-pin-fin: Numerical study. International Communications in Heat and Mass Transfer. 2020;**112**(February):104455. DOI: 10.1016/j.icheatmasstransfer.2019. 104455

[44] Spence, R. and Amaral-Teixeira, J., 2009.A CFD parametric study of geometrical variations on the pressure pulsations and performance characteristics of a centrifugal pump. Computers and Fluids, 38(6), pp.1243-1257.

[45] Feng J, Benra FK, Dohmen HJ. Application of different turbulence models in unsteady flow simulations of a radial diffuser pump. For schIngenieurwes. 2010;74(3):123-133

[46] Li WG. Effect of exit blade angle, viscosity and roughness in centrifugal pumps investigated by CFD computation. Task Quarterly. 2011;5:21-41

[47] Hedi ML, Hatema K, Ridha Z. Numerical analysis of the flow through in centrifugal pumps. International Journal of Thermal Technologies. 2012; 2(4):216-221

[48] Lei T, Baoshan Z, Shuliang C, Hao B, Yuming W. Influence of blade wrap angle on centrifugal pump performance by numerical and experimental study. Chinese Journal of Mechanical Engineering. 2014;27(1):171-177

[49] Bellary, S.A.I. and Samad, A., 2015-b. Numerical analysis of centrifugal impeller for different viscous liquids. International Journal of Fluid Machinery and Systems, 8(1), pp.36-45.

[50] Yong, W., Lin, L.H., Qi, Y.S., Gao, T.M. and Kai, W., 2009. Prediction research on cavitation performance for centrifugal pumps. International Conference on Intelligent Computing and Intelligent Systems, IEEE, 1(1), pp. 137-140.

[51] Proudman I. The generation of noise by isotropic turbulence. Proceedings of the Royal Society A: Mathematical, Physical and Engineering Sciences. 1952;**214**

[52] Jaiswal AK, Rehman AU, Paul AR, Jain A. Detection of cavitation through acoustic generation in centrifugal pump impeller. Journal of Applied Fluid Mechanics. 2019;**12**(4)

[53] Mishra AK, Rehman AU, Paul AR, Jain A. Performance analysis of an impeller of a centrifugal pump using OpenFOAM. In: Proceedings of 6th International and 43rd National Conference on Fluid Mechanics and Fluid Power (FMFP-2016) 15-17 Dec. 2016. 2016

[54] Rehman AU, Paul AR, Jain A. Performance analysis and cavitation prediction of centrifugal pump using various working fluids. Recent Patents on Mechanical Engineering. 2019;**12**(3): 227-239. DOI: 10.2174/ 2212797612666190619161711

[55] Rehman AU, Srivastava S, Jain A, Paul AR, Mishra R. Effect of blade trailing edge angle on the performance of a centrifugal pump impeller. In: Proceedings of COMADEM-2015. 2015

[56] Rehman AU, Yadav P, Paul AR, Jain A. Effect of number of blades on performance of a centrifugal pump. In: Proceedings of 6th International and 43rd National Conference on Fluid Mechanics and Fluid Power (FMFP-2016) 15-17 Dec. 2016, MNNIT Allahabad (India). 2016

[57] Kalyan DK, Rehman AU, Paul AR, Jain A. Computational flow analysis through a double-suction impeller of a centrifugal pump. In: Proceedings of 40th National Conference on Fluid Mechanics & Fluid Power (FMFP-2013), 12-14 December 2013, NIT Hamirpur, Himachal Pradesh. 2013

[58] Jaiswal AK, Siddique H, Paul AR, Samad A. Surrogate based design optimization of a centrifugal pump impeller. In: Engineering Optimization. 2021

A CFD Porous Materials Model to Test Soil Enriched with Nanostructured Zeolite Using ANSYS-Fluent⁽™⁾

Diana Barraza-Jiménez, Sandra Iliana Torres-Herrera,
Patricia Ponce Peña, Carlos Omar Ríos-Orozco,
Adolfo Padilla Mendiola, Elva Marcela Coria Quiñones,
Raúl Armando Olvera Corral, Sayda Dinorah Coria Quiñones
and Manuel Alberto Flores-Hidalgo

Abstract

Soil health is a great concern worldwide due to the huge variety of pollutants and human activities that may cause damage. There are different ways to remediate and make a better use of soil and a choice may be using zeolite in activities like gardening, farming, environment amending, among others. In this work is proposed a model to simulate how mixing zeolite with soil may be beneficial in different ways, we are especially interested in interactions of mixed soil-zeolite with water. This model is based in different flow regimes where water interacts with two layers formed by nanostructured zeolite and soil in a vertical arrangement. The analysis is approached as a bi-layer porous material model resolved by using the mathematical model implemented in ANSYS-Fluent. Such model uses a multi-fluid granular model to describe the flow behavior of a fluid–solid mixture where all the available interphase exchange coefficient models are empirically based. Despite the great capabilities of numerical simulation tools, it is known that at present time, the literature lacks a generalized formulation specific to resolve this kind of phenomena where a porous media is analyzed. This model is developed to obtain a systematic methodology to test nanomaterials with porous features produced in our laboratory which is the next step for near future work within our research group.

Keywords: porous, CFD, ANSYS-Fluent®, soil, nanostructured zeolite

1. Introduction

A porous material is a complex structure consisting of a compact phase (usually solid) and some void space, which relates directly with the term porosity. The literature describes a porous medium as a region in space comprising of at least two homogeneous material constituents, presenting identifiable interfaces between them in a resolution level, with at least one of its constituents remaining fixed or slightly deformable [1]. Among porous materials, soil and zeolite are interesting because they are perfectly aligned with the definition for porous materials.

The aim of this work is search for options to improve soil health since it is a great concern worldwide due to the huge variety of pollutants and anthropogenic activities that may cause damage. Zeolite is an option to amend soil in activities like gardening, farming, environment amending, among others, it is reported in the literature as a suitable material for sustainable chemistry. Mixing zeolite in soil may be beneficial in different ways, we are especially interested in interactions of mixed soil-zeolite with water. In this work, a model is developed to obtain a systematic methodology to test nanomaterials with porous features produced in our laboratory which is the next step for near future work within our research group. This model is based in different flow regimes where water interacts with two layers formed by nanostructured zeolite and soil in a vertical arrangement. The analysis is approached as a bi-layer porous material model resolved using the mathematical model implemented in ANSYS-Fluent.

2. Fundamental concepts

In this section a brief set of fundamental concepts are displayed to put the reader in context with the topics within this research work as described next.

2.1 Porous materials

The word *porous* describes a structure with a compact phase (usually solid) and some void (empty) space. Any solid material containing cavities, channels or interstices may be considered porous. Concepts like "pores", "cavities', among others, are features for a porous material and according to them, such material may be characterized. The reader is referred to the literature on porous materials for more information on the correct use of related terms [1–5]. Our recommended definition for the context of the present work is "*A porous medium is a region in space comprising of at least two homogeneous material constituents, presenting identifiable interfaces between them in a resolution level, with at least one of the constituents remaining fixed or slightly deformable*" [2]. Measurements are important to characterize porous materials, one may be interested in pore size range, performance at different levels of compression and/or temperature, performance when interacting with different liquids, repeatability, structures, corrugated pores, etc. [1] Even though, porous materials latest breakthrough was reported over a decade ago, there are still plenty of interesting applications that use this type of materials [4] and new techniques to prepare them. Also, over the last ten years or so, literature highlights include synthesis, applications, hierarchically structured porous materials [1–6], among others.

Porous materials are defined as elements/compounds that contain a porous structure consisting of interconnected pores on different length scales from micro- (<2 nm), meso- (2–50 nm) to macropores (>50 nm). Micro- and mesopores may provide size and shape selectivity for guest molecules, enhancing the host–guest interactions. Alternatively, macropores can considerably favor diffusion to and accessibility of active sites by guest molecules, which is particularly important for the diffusion of large molecules or in viscous systems. Emphasis in porous size is an important work trend among scientists and technologists due to the wide range of possibilities regarding applications of porous materials based in pore size. One of the more interesting porous materials is zeolites which are included within this study [5–7].

2.2 Zeolites

Zeolites were found in 1756 and since then their use has spread out in chemical industries for catalysis, adsorption, separation, and a great variety of other applications. 35,232 patents with the title including "zeolite*" are documented by Derwent Innovations Index as of January 2, 2020 and around 30,271 publications with "zeolite*" in their title are recorded by the Web of Science Core Collection in the same date. Although there is a lot of work and advancement in the science and technology related to zeolites, fundamental research on them and their applications have a great deal of relevancy [5].

Zeolites in its natural mineral presentation are found in several parts of the world but most zeolites used are produced by synthesis [8, 9]. Differences between natural and synthetic zeolites include: 1) Synthetics are obtained from chemicals and naturals are processed mines, 2) Synthetic zeolites silica to alumina ratio is 1 to 1 and natural clinoptilolite zeolites is 5 to 1 ratio, 3) clinoptilolite zeolite do not break down in mildly acid environment, synthetic zeolites do break. Natural zeolite structure has more acid resistant silica to keep its structure together [9].

Zeolite is a microporous (<2 nm) material comprising crystalline aluminosilicate with various structures [10, 11]. Over 200 types of zeolites have been reported [12] with pore diameters between 0.25 and 1 nm [13] and possess good selectivity properties [14–16]. In catalytic applications, zeolite framework structure is an assembly made of AlO_4 and SiO_4 tetrahedra able to provide Brønsted and Lewis acid sites inside the micropore [17–20]. For example, Brønsted acid sites in synthetic zeolites, such as zeolite Y and ZSM-5, are responsible for the catalytic cracking reaction in oil refinery [21].

Hierarchical porous zeolite addresses issues with porous size. Under its perspective, there are three types of porosity according to pore size, micropore (<2 nm), mesopore (2–50 nm), and macropore (>50 nm) [22]. Zeolites may be considered a family of crystalline aluminosilicates consisting of orderly distributed molecular sized nanopores. Their structure benefits adsorption of guest molecules with specific sizes and shapes or separation processes for liquid or gas mixtures as molecular sieves [23, 24]. In addition, zeolites with guest species, coupled with acid or metal sites, enables shape-selective catalysis [25–27]. Zeolites are considered the most important solid catalysts in petrochemical industries [28–31]. Zeolitic materials are also promising in a wide variety of applications, including renewable energy and environmental improvement [32].

Properties of zeolites are directly related with their nanoporous framework structures, so TO_4 tetrahedra ("T" denotes tetrahedrally coordinated Si, Al, P, etc.) is fundamental [33]. According to the literature [22], 235 types of zeolite frameworks have been discovered [12], however, there is still a high demand for improved zeolitic materials with new structures and superior functions. In addition, new technology trends are giving a new impulse to zeolite research, such is the case of nanotechnology where nanostructured zeolite or interactions of zeolite with nanostructured materials have captured the interest of researchers. Soil may be counted among the more interesting interactions with zeolite.

2.3 Soil

The Soil Science Society of America has published two definitions for soil. One is *"The unconsolidated mineral or organic material on the immediate surface of the earth that serves as a natural medium for the growth of land plants."* The second definition may be more inclusive and says soil is *"The unconsolidated mineral or organic matter on the surface of the earth that has been subjected to and shows the effects of genetic and* environmental factors of: climate (including water and temperature effects) and macro- and microorganisms, conditioned by relief, acting on parent material over a period of time" [34, 35].

A soil detailed definition depends upon physical, chemical, biological, and morphological properties, and characteristics. Their effect on soil management decisions is critical in any case the soil is to be used in either crop production, in an urban setting, or for roads, dams, waste disposal, and other uses [35].

Soil is a porous media at the land surface formed by weathering processes mediated by biological, geological, and hydrological phenomena. Soil is different than weathered rock because it shows a vertical stratification (the soil horizons) that has been produced by the influence of percolating water and living organisms. From a chemistry perspective, soils are open, multicomponent, biogeochemical systems containing solids, liquids, and gases. Open systems mean soils exchange matter and energy with the surrounding atmosphere, biosphere, and hydrosphere. Such exchange is highly variable, but it is the essential flux that cause the development of soil profiles and the patterns of soil quality [36].

Generally, soil is formed by fragmented and chemically weathered rock which includes sand, silt, and clay separates, and contains humus (partially decomposed organic matter). Soil diversity is huge, because of the different regional circumstances, it varies considerably. If properties of soil are known, it may be effectively managed and succeed at a specific use or purpose.

The major elements in soils exceed a concentration of 100 mg-kg^{-1}, all others are known as trace elements. According to multiple reports, the major elements include O, Si, Al, Fe, C, K, Ca, Na, Mg, Ti, N, S, Ba, Mn, P, and perhaps Sr and Zr, in decreasing order of concentration. The major elements C, N, P, and S also are macronutrients, so they are critical to life cycles and may be absorbed by organisms in significant amounts [36].

2.4 Water and drought

Climate change is affecting the way we live without a doubt. For example, El Nino and La Nina are climate patterns in the Pacific Ocean that affect weather worldwide [10]. In Mexico, these and other climate related phenomena are responsible for intensified drought in a great part of the country. In the northern part of Mexico for this year (2021) the forecast indicates there will be 20–30% water availability for the different activities if compared to last year [11]. Porous materials may be a feasible option for gardens and crop soil to keep humidity for longer periods of time. Then, porous materials and specially zeolite, are interesting materials for studying their interaction with soil and water.

3. Methodology

In this work is used the code ANSYS-Fluent® [37, 38] and all CFD methodologies presented are embedded in this program. Like most CFD codes, ANSYS contains three main elements: (a) a preprocessor, (b) a solver, and (c) a postprocessor, the role of each one will be described briefly in the next sections.

ANSYS Fluent solves conservation equations for mass and momentum. For flows involving heat transfer or compressibility, an additional equation for energy conservation is solved. Additional transport equations are solved when the flow has other features such as transport species, chemical reactions, turbulence, etc. Since conservation equations are widely known, we will present only the simplified version and will focus in describing the porous media approach briefly.

The equation for conservation of mass, or continuity equation, can be written as follows:

$$\frac{\partial \rho}{\partial t} + \nabla \bullet \left(\rho \vec{v} \right) = S_m \tag{1}$$

Eq. (1) is the general form of the mass conservation equation and is valid for incompressible as well as compressible flows. The source S_m is the mass added to the continuous phase from the dispersed second phase (for example, due to vaporization of liquid droplets) and any user-defined sources.

Conservation of momentum in an inertial (non-accelerating) reference frame is described by the next equation [39]:

$$\frac{\partial}{\partial} \left(\rho \vec{v} \right) + \nabla \bullet \left(\rho \vec{v} \vec{v} \right) = -\nabla p + \nabla \bullet \left(\overline{\overline{\tau}} \right) + \rho \vec{g} + \vec{F} \tag{2}$$

where p is the static pressure, $\overline{\overline{\tau}}$ is the stress tensor (described below), $\rho \vec{g}$ is the gravitational body force and $\vec{F}$ represent external body forces (for example, those arising from interaction with the dispersed phase). $\vec{F}$ also contains other model-dependent source terms such as porous-media and user-defined sources.

3.1 Geometry and meshing

Geometry and meshing are part of the preprocessing phase to resolve a computational fluid dynamics problem. The definition of the key features of our model starts with the idea of simulating a water flow through a porous zone formed by a thin layer of zeolite applied over a layer of soil. This model consists of a vertical arrangement of a packed bed like porous zone formed by the two layers, both contained in a transparent pipe with a water flow from top to bottom applied by gravity. PTC-CREO [40] was used to develop the 3D CAD model needed so CAE software may be enabled to carry on with the CFD simulation. SpaceClaim is a module within ANSYS used to prepare geometries for CFD calculations [37] and, it was used to extract the fluid domain for the meshing procedures. The module ANSYS meshing was used to carry on with the meshing procedure of the fluid domain.

3.2 Porous media model

The porous media model incorporated in ANSYS-Fluent can be used in a wide variety of single phase and multiphase problems, for example, flow through packed beds, filter papers, perforated plates, flow distributors, and others.

In this model, a cell zone is selected as the porous media where ANSYS methodology is applied by means of user inputs and the Momentum Equations for Porous Media, for further information the reader is referred to the ANSYS-Fluent manual [38].

3.3 Limitations and assumptions of the porous media model

The porous media model incorporates an empirically determined flow resistance in a region of your model defined as "porous". In essence, the porous media model adds a momentum sink in the governing momentum Equations [38]. The model would represent a porous zone without a detailed exact model of the porosity within the materials at microscale, in other words, the porous zone will be represented qualitatively and will be resolved with equations empirically defined to do so as explained briefly in the next paragraphs.

3.4 Momentum equations for porous media

The porous media models for single phase flows and multiphase flows use the Superficial Velocity Porous Formulation as the default. ANSYS Fluent calculates the superficial phase or mixture velocities based on the volumetric flow rate in a porous region.

Porous media are modeled by the addition of a momentum source term to the standard fluid flow equations. The source term is composed of two parts: a viscous loss term (Darcy's equation first term on the right-hand side, and an inertial loss term (Darcy's equation second term on the right-hand side), as shown next:

$$S_i = -\left(\sum_{j=1}^{3} D_{ij}\mu v_j + \sum_{j=1}^{3} C_{ij}\frac{1}{2}\rho|v|v_j\right) \tag{3}$$

where S_i is the source term for the i-th (x, y, or z) momentum equation, $|v|$ is the magnitude of the velocity and D and C are prescribed matrices. This momentum sink contributes to the pressure gradient in the porous cell, creating a pressure drop that is proportional to the fluid velocity (or velocity squared) in the cell which enables a viable calculation route without the need of microscale porosity details.

To recover the case of simple homogeneous porous media

$$S_i = -\left(\frac{\mu}{\alpha}v_i + C_2\frac{1}{2}\rho|v|v_j\right) \tag{4}$$

where α is the permeability and C_2 is the inertial resistance factor, simply specify D and C as diagonal matrices with $1/\alpha$ and C_2, respectively, on the diagonals (and zero for the other elements).

ANSYS Fluent also allows the source term to be modeled as a power law of the velocity magnitude [38]:

$$S_i = -C_0|v|^{C_1} = -C_0|v|^{(C_1-1)}v_i \tag{5}$$

where C_0 and C_1 are user-defined empirical coefficients.

Important

In the power-law model, the pressure drop is isotropic and the units for C_0 are SI but we will not enter any further in the mathematical model supporting this idea and will refer the reader to ANSYS manuals and proper references contained in there to develop this model in part or in whole at the reader convenience.

3.5 Darcy law and Darcy-Forchhimer

In laminar flows through porous media, the pressure drop is typically proportional to velocity and the constant can be considered zero. Ignoring convective acceleration and diffusion, the porous media model then reduces to Darcy's Law [38]:

$$\nabla p = -\frac{\mu}{\alpha}\vec{v} \tag{6}$$

Pressure drop is computed in ANSYS Fluent for each one of the three (x, y, z) coordinate directions within the porous region according to:

$$\Delta p_x = \sum_{j=1}^{3} \frac{\mu}{\alpha} v_j \Delta n_x$$

$$\Delta p_y = \sum_{j=1}^{3} \frac{\mu}{\alpha} v_j \Delta n_y \qquad (7)$$

$$\Delta p_z = \sum_{j=1}^{3} \frac{\mu}{\alpha} v_j \Delta n_z$$

where $1/\alpha_{ij}$ are the entries in the matrix D in Eq. (3), v_i are the velocity components in the x, y, and z directions, and Δn_x, Δn_y, and Δn_z are the thicknesses of the medium in the x, y, and z directions.

Here, the thickness of the medium (Δn_x, Δn_y, or Δn_z) is the actual thickness of the porous region in our model. Therefore, if the thicknesses used in the model differ from the actual thicknesses, adjustments may be needed in the inputs for $1/\alpha_{ij}$.

Calculations for laminar flow regime models were based in Darcy law. For calculations under turbulent flow regime Darcy-Forchhimer is the mathematical model used by ANSYS.

3.6 Processing

This work was processed using ANSYS Fluent®. The pressure–velocity coupling scheme controls the way pressure and velocity are updated when the pressure-based solver is used. The scheme can be either segregated (pressure and velocity are updated sequentially) or coupled (pressure and velocity are updated simulta-neously) [38]. The scheme used in this work for pressure–velocity coupling is SIMPLE. For spatial discretization we use least squares cell based for gradient, PRESTO! for pressure and second order upwind for momentum. This set up was successful to treat single layer and double layer porous media models and convergence was reached with few to moderate number of iterations.

3.7 Postprocessing

The results module provided by ANSYS® was used to visualize code/numerical results, the data may be presented in different ways to facilitate the numerical analysis. The figures and graphs were generated from the numerical sheet produced within ANSYS-Fluent. These may include domain geometry and grid display, vec-tor plots, line, and shaded contour plots, 2D and 3D surface plots, particle tracking, and view in perspective (translation, rotation, scaling, etc.), and few hand-made numerical computations.

4. Results and discussion

This work was developed using computational fluid dynamics (CFD) as implemented in ANSYS-Fluent. The calculation processes are explained in the next paragraphs.

4.1 Geometry and meshing

The geometries and assemblies initially were developed using CAD programs, but they can be designed either way in the geometry module within ANSYS, which is called SpaceClaim. A CAD program enables further development and a highly detailed design, which is interesting for complex developments. For the scope of this work, SpaceClaim was used to prepare the geometry for CFD. **Figure 1** shows the geometry preparation in different steps up to the meshing generation.

Meshing procedure was carried on ANSYS meshing module, since the geometry is a simple cylinder, the discretization process was easily resolved. The meshing model was carried on systematically with a different number of elements, from a

Figure 1.
Geometry used to simulate a single layer of zeolite exposed to water flow from top to bottom. (a) CAD geometry for zeolite single layer interacting with water, (b) CAD geometry of fluid and porous zone without pipe and covers. (c) Fluid domain prepared in ANSYS for CFD simulation, (d) meshed fluid domain.

rough mesh to a finer mesh in search of the more efficient model. Overall, under 100 thousand elements was considered a rough mesh, up to 500 thousand elements is medium and over that number of elements is considered a fine mesh. The models used in 3D demonstrated a nice performance during the convergence trials (**Figure 2**), however, if a further simplification is found, it should be considered.

Therefore, 2D models were developed to improve efficiency in our calculations. The 2D model worked very well and improved efficiency so we decided to present the results generated with these models.

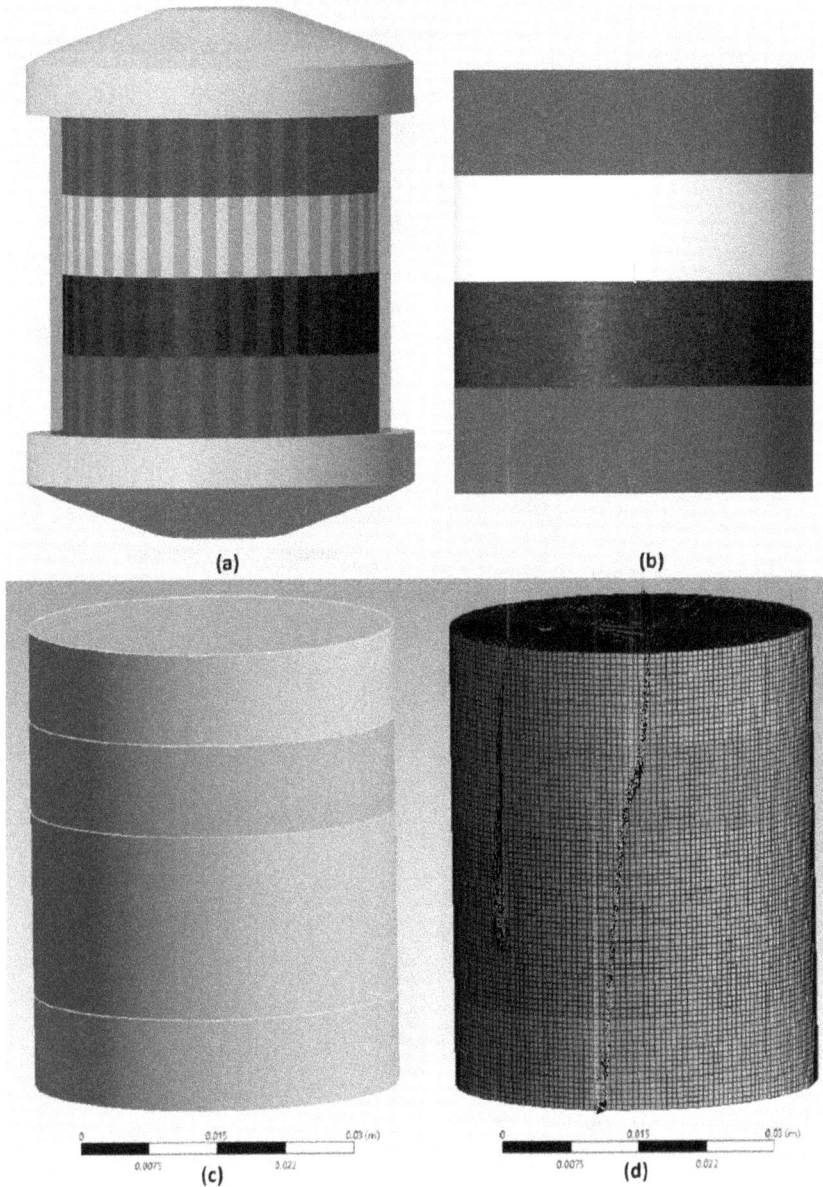

Figure 2.
Geometry used to simulate a layer of zeolite over a layer of soil exposed to water flow from top to bottom. (a) CAD geometry for zeolite-soil porous layers interacting with water, (b) CAD geometry of fluid and porous zone without pipe and covers. (c) Fluid domain prepared in ANSYS for CFD simulation, (d) meshed fluid domain.

Overall, the best results were obtained with simplified models using a 2D geometry representative of the proposed systems with single and double porous media layer, an example of 2D model geometry used is shown in **Figure 3**. The figure illustrates a slice of the interacting materials stacked from top to bottom with a first layer of water on top, followed by two layers composed with porous materials ordered in zeolite placed over soil and at the bottom more water. **Figure 3(a)** was obtained from ANSYS SpaceClaim where it may be optional to label each layer but also labels may be added in the meshing module. The geometry was simplified to a basic shape, as can be observed in **Figure 3(b)**. Thus, the model was easily

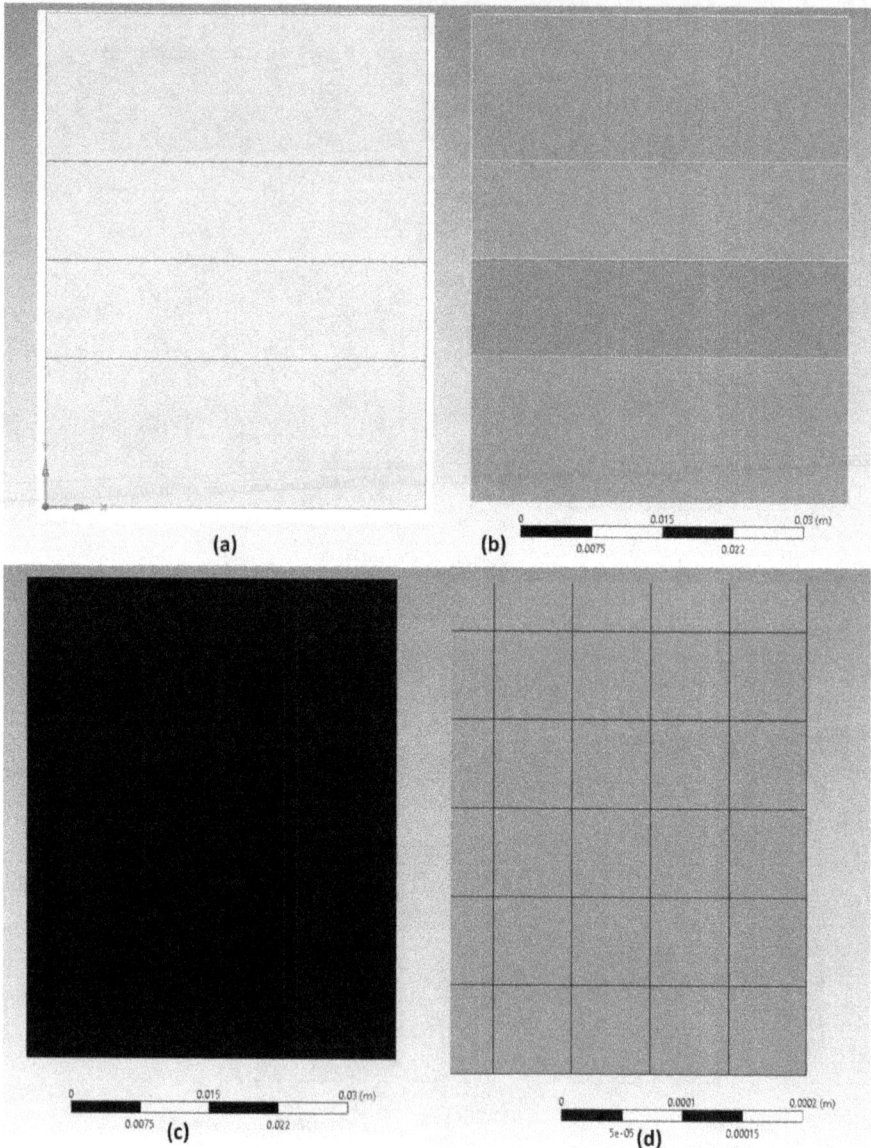

Figure 3.
Simplification to a 2D model of the double layer with soil zeolite on top of soil as loaded in ANSYS for meshing procedures and mesh obtained (see scale for length dimensions under the model image).

discretized by using ANSYS meshing module with an average size element of 6.5×10^{-5} m (0.065 mm) which in total added up to 473,550 elements. The discretization results may be observed in **Figure 3(c)** and **(d)**.

4.2 Solution

Convergence trials were carried on with different geometries and resolution models. The best choice was selected based in efficiency and 2D models were selected over 3D. The geometries proposed were designed as simple models, the idea is a simple pipe containing a packed bed formed with one or two porous materials layers.

These geometries were used in calculations under laminar and turbulent flow regime set up subject to boundary conditions. Calculations with double layer and under both flow regimes are included in the following sections for a more detailed discussion of results.

4.3 Laminar flow regime

Two different set of calculations are presented in the next paragraphs, the first one is based in results obtained from laminar flow regime models. Results for pressure calculations are presented in **Figure 4**. These calculations required an input velocity with different values relatively low to obtain laminar flow through a double layer porous zone built with zeolite and soil. Velocity values used in this section are $v_1 = 0.005$ m/s, $v_2 = 0.01$ m/s, $v_3 = 0.02$ m/s, $v_4 = 0.03$ m/s, and $v_5 = 0.04$ m/s.

Pressure effects are displayed in **Figure 4**, to understand pressure-drop in a layer-by-layer contour plot that illustrates water flow moving through zeolite and soil layers modeled as porous media within ANSYS-Fluent.

The higher the input velocity, the higher pressure is required to make the flow pass through the porous media, for specific pressure values a scale in pascals is shown by the side of each simulation to help interpret the contours color in the image.

In **Figure 5** are presented contour plots of velocity to illustrate how water is applied gradually into the model. Water is applied using an input velocity with low values to keep the flow under laminar regime in y-axis negative direction (downwards). Velocity decreases as the flow advances through the pipe and porous zone represented by the two layers simulating zeolite and soil. Each velocity contour plot includes a scale with velocity values in meters per second to facilitate the interpretation of each color included in the contour plot. For a better understanding of pressure drop, a graph showing pressure drop profile was generated based in results for laminar flow regime computations as displayed in **Figure 6**. This profile is built as a scatter plot using y-axis or height in the model as the *x-coordinate* or abscissa and, pressure drop was represented in the *y-ordinate*. The scatter plot displays an overall view of the pressure drop as y-axis values change through the pipe and porous zone. To analyze further the effects on pressure drop for laminar flow regime simulations, pressure values at different locations were calculated, specifically, at the inlet and outlet for each layer of porous media zone (boundaries). Also, pressure drop for each layer was calculated by finding the difference between pressure values at layers inlet and outlet (this difference will be referred to as delta pressure values). **Table 1** contains pressure and delta pressure calculations numerical results.

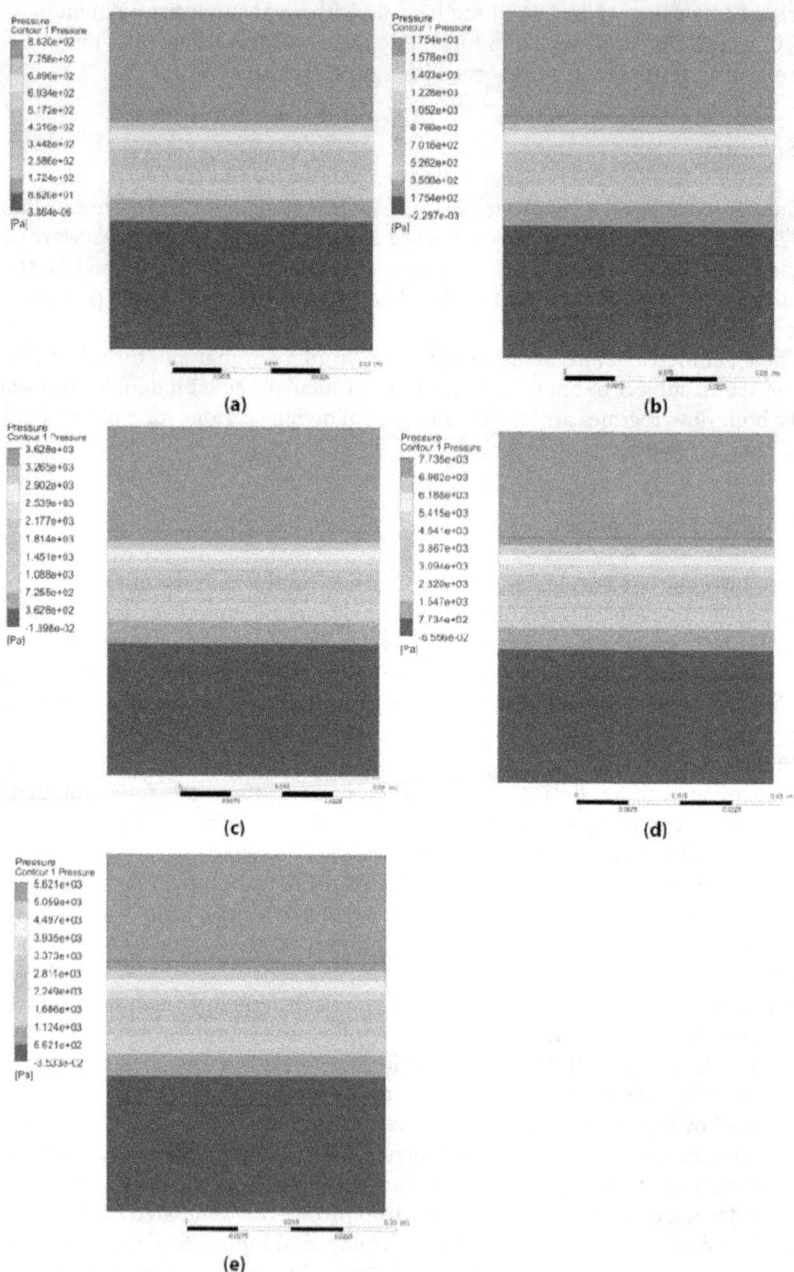

Figure 4.
Contour plots corresponding to results for pressure from laminar flow regime calculations using a double layer model porous zone. (a) $v_1 = 0.005$ m/s, (b) $v_2 = 0.01$ m/s. (c) $v_3 = 0.02$ m/s, (d) $v_4 = 0.03$ m/s, (e) $v_5 = 0.04$ m/s.

4.4 Turbulent flow regime

Results for pressure calculations obtained from turbulent flow model simulations are presented in **Figure 7**. These calculations required an input velocity with different values to obtain turbulent flow through our model with a double layer porous zone built with zeolite and soil.

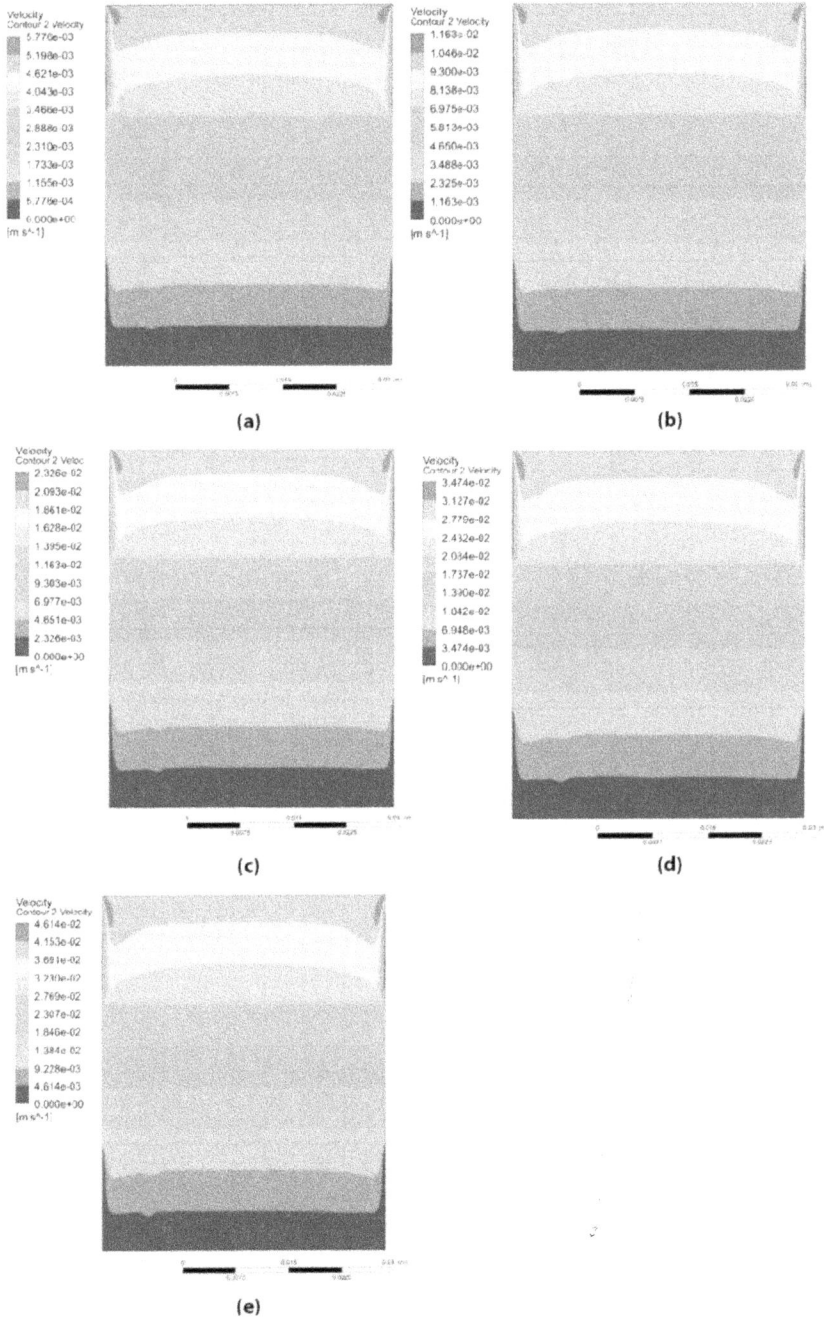

Figure 5.
Contour plots corresponding to results for velocity in y-direction from laminar flow regime calculations using a double layer model porous zone. (a) v_1 = 0.005 m/s, (b) v_2 = 0.01 m/s. (c) v_3 = 0.02 m/s, (d) v_4 = 0.03 m/s, (e) v_5 = 0.04 m/s.

Velocity values used in this section are v_1 = 0.04 m/s, v_2 = 0.05 m/s, v_3 = 0.1 m/s, v_4 = 0.2 m/s, v_5 = 0.3 m/s, v_6 = 0.4 m/s, v_7 = 0.5 m/s. Pressure effects as displayed in **Figure 7** represent pressure drop for our models under turbulent flow regime in a

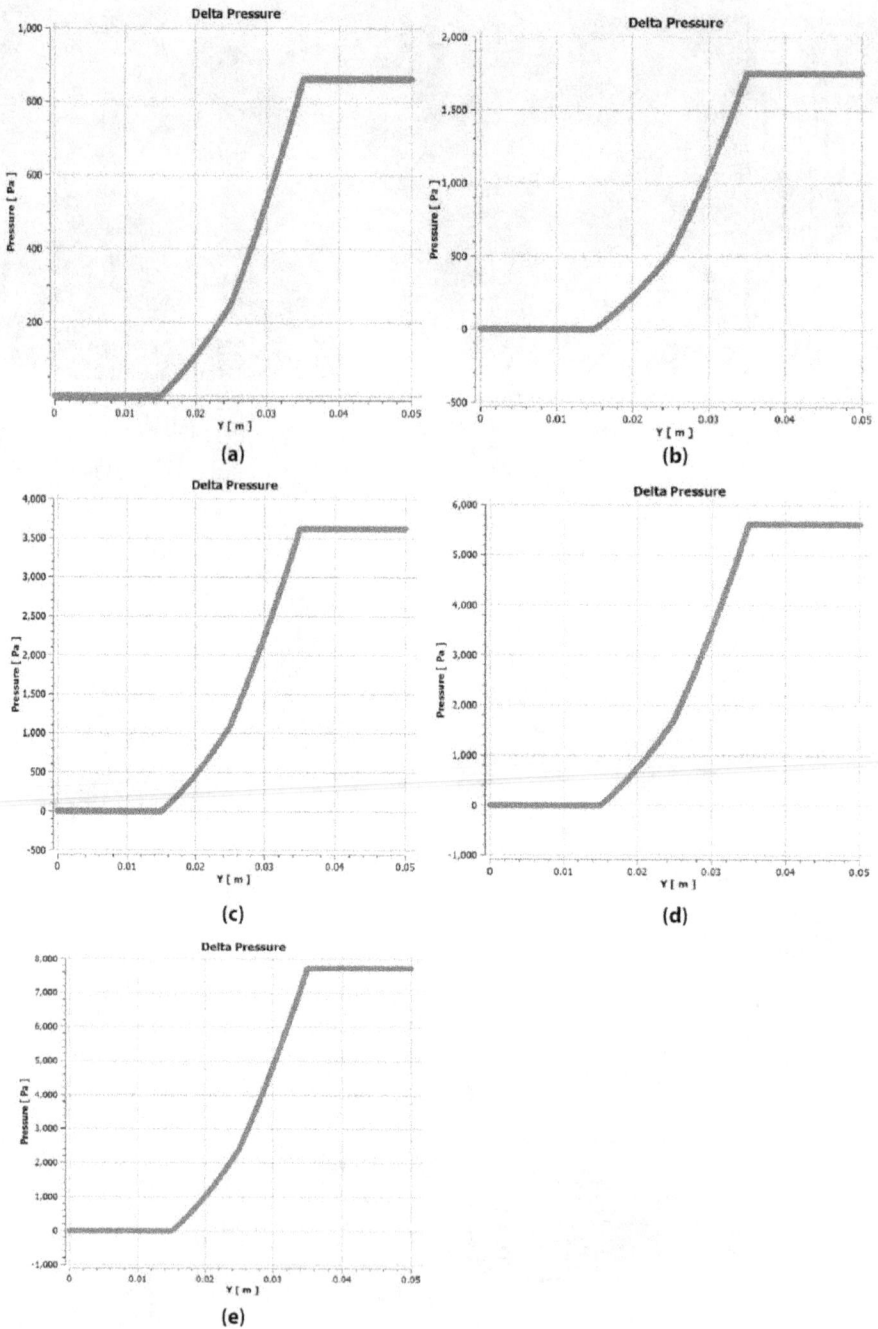

Figure 6.
Graph showing pressure drop results for laminar flow regime calculations using a double layer model porous zone. (a) $v_1 = 0.005$ m/s, (b) $v_2 = 0.01$ m/s. (c) $v_3 = 0.02$ m/s, (d) $v_4 = 0.03$ m/s, (e) $v_5 = 0.04$ m/s.

layer-by-layer contour plot to illustrate water flow through zeolite and soil layers modeled as porous media within ANSYS-Fluent.

The higher the input velocity, the higher pressure is required to make the flow through the porous media, for specific pressure values a scale in pascals is shown by

v (m/s)	$P_{inlet\text{-}zeol}$ (Pa)	$P_{outlet\text{-}zeol}$ (Pa)	$P_{inlet\text{-}soil}$ (Pa)	$P_{outlet\text{-}soil}$ (Pa)	ΔP_{zeol} (Pa)	ΔP_{soil} (Pa)	ΔP_{Total} (Pa)
0.005	862	603.4	258.6	86.2	258.6	172.4	775.8
0.01	1754	701.6	526.2	175.4	1052.4	350.8	1578.6
0.02	3628	1451	1088	362.9	2177	725.1	3265.1
0.03	7735	3094	2320	773.4	4641	1546.6	6961.6
0.04	5621	2249	1686	562.1	3372	1123.9	5058.9

Table 1.
Laminar flow regime pressure-drop numerical results at the boundaries between different layers (zeolite over soil) to analyze flow through porous zone.

the side of each simulation to help interpret the contours color in the image. In comparison with laminar flow, water flow velocity and pressure present higher values.

In **Figure 8** are presented contour plots of velocity to illustrate how water flows through the porous zone. Water is applied using an input velocity with low values just enough to keep the flow as turbulent with a direction in y-axis with or without negative sign (downwards).

Velocity decreases as the flow advances through the porous zone represented by the two layers simulating zeolite and soil. Each velocity contour plot includes a scale with velocity values in meters per second to facilitate the interpretation of each color included in the contour plot.

For a better understanding of pressure drop, a graph showing pressure drop profile was generated for turbulent flow calculations as displayed in **Figure 9**. Similarly, as it was done with laminar flow, the profile is built with a scatter plot using y-axis or height in our model as the *x-coordinate* and pressure drop is represented in the *y-ordinate*. Also, to analyze further the effects on pressure drop for laminar flow regime simulations, pressure values at the boundaries for each layer of the porous media zone. Pressure drop for each layer in the turbulent model was calculated by finding the difference between pressure values at layers considering an inlet and an outlet. **Table 2** contains pressure and delta pressure numerical results.

4.5 Effects on zeolite and soil layers

Porous zone flow is simulated as a region that presents resistance to the fluid flow. When water is introduced in the system each layer representing a porous material presents a difficulty to allow flow through which can be measured with the pressure drop calculated on those areas. Due to its properties, zeolite layer presents the higher pressure drop values. Zeolite and soil material parameters to represent materials properties used within this work are based in textbook values [1, 8, 35] and can be modified as required depending on the specific properties of the materials that need to be simulated. The input velocity is also important regarding how pressure drop displays its profile and relative values, in general, the higher the input velocity value, the higher the pressure drop in the porous zone areas. Such effect occurs in laminar flow and turbulent flow. However, pressure drop may be higher in turbulent flow due to velocity input values are higher too. This model may be useful for future developments where the porous materials properties are modified or when one needs further studies related to water distribution in the system.

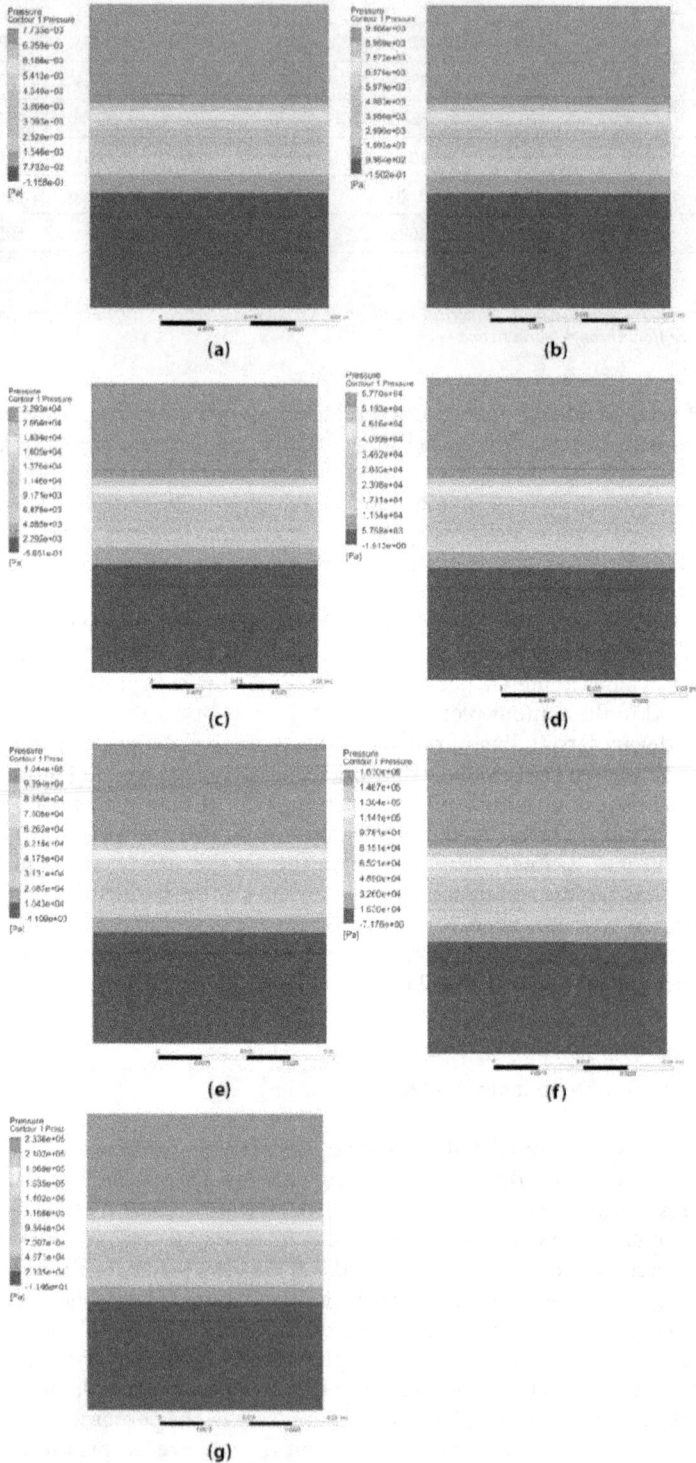

Figure 7.
Contour plots for pressure obtained from turbulent flow regime numerical results corresponding to calculation with different velocity inputs using a double layer porous zone, the velocity values used were: (a) v₁ = 0.04 m/s, (b) v₂ = 0.05 m/s. (c) v₃ = 0.1 m/s, (d) v₄ = 0.2 m/s, (e) v₅ = 0.3 m/s, (f) v₆ = 0.4 m/s, (g) v₇ = 0.5 m/s.

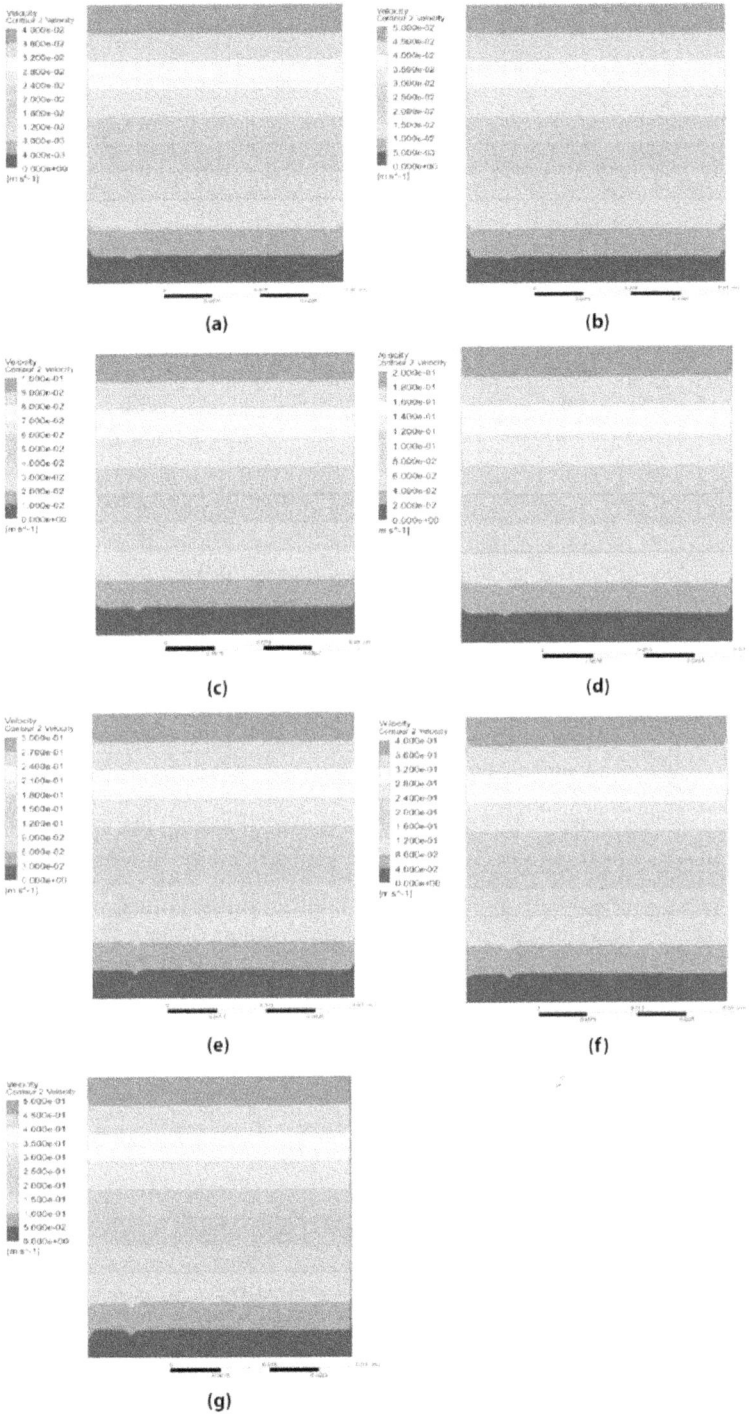

Figure 8.
Contour plots for velocity in y-axis obtained from turbulent flow regime numerical results corresponding to calculations with different velocity inputs using a double layer porous zone, the velocity values used were: (a) v_1 = 0.04 m/s, *(b)* v_2 = 0.05 m/s. *(c)* v_3 = 0.1 m/s, *(d)* v_4 = 0.2 m/s, *(e)* v_5 = 0.3 m/s, *(f)* v_6 = 0.4 m/s, *(g)* v_7 = 0.5 m/s.

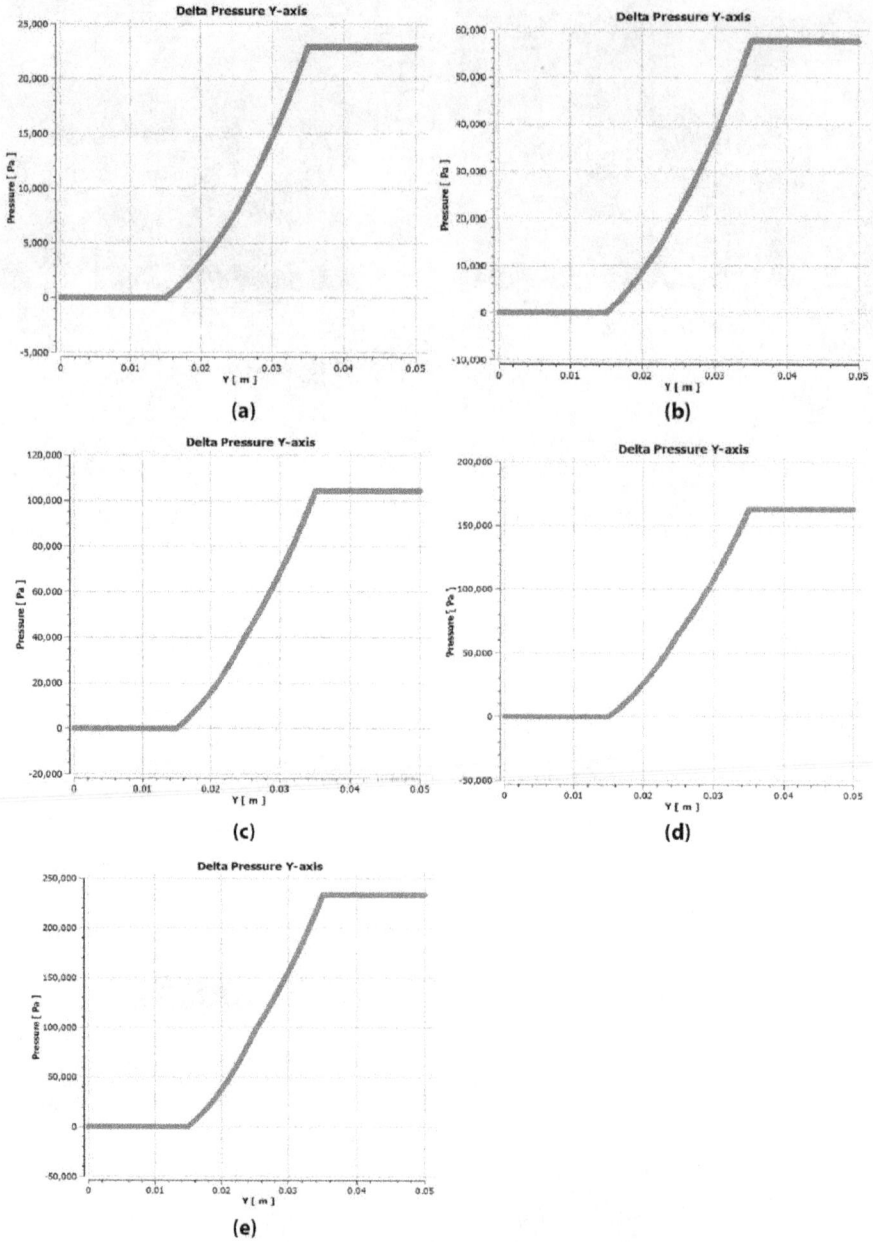

Figure 9.
Graph showing pressure drop results for turbulent flow regime calculations using a double layer model porous zone. (a) $v_1 = 0.04$ *m/s, (b)* $v_2 = 0.05$ *m/s. (c)* $v_3 = 0.1$ *m/s, (d)* $v_4 = 0.2$ *m/s, (e)* $v_5 = 0.3$ *m/s, (f)* $v_6 = 0.4$ *m/s, (g)* $v_7 = 0.5$ *m/s.*

5. Conclusions

Computational fluid dynamics (CFD) is used as a powerful tool to analyze multi-physics problems in a wide variety of applications. To analyze porous materials ANSYS-Fluent offers an interesting scheme that enables the study of a fluid through a porous material. A bi-layer model was built to represent a layer of zeolite placed

v (m/s)	$P_{inlet\text{-}zeol}$ (Pa)	$P_{outlet\text{-}zeol}$ (Pa)	$P_{inlet\text{-}soil}$ (Pa)	$P_{outlet\text{-}soil}$ (Pa)	ΔP_{zeol} (Pa)	ΔP_{Soil} (Pa)	ΔP_{Total} (Pa)
0.04	7733	3093	2320	773.2	4640	1546.8	6959.8
0.05	9966	3986	2990	996.4	5980	1993.6	8969.6
0.1	22930	9171	6878	2292	13759	4586	20638
0.2	57700	28850	23080	5768	28850	17312	51932
0.3	104400	52180	31310	10430	52220	20880	93970
0.4	163000	81510	65210	16300	81490	48910	146700
0.5	233600	116800	93440	23350	116800	70090	210250

Table 2.
Turbulent flow regime pressure-drop calculations at the boundaries between different layers (zeolite over soil) to analyze flow through porous zone.

over a layer of soil and both interacting with a water flow. Laminar and turbulent flow regimes were analyzed successfully with the approach proposed which represents an attempt to systematically analyze different nanostructured zeolites interacting with different soil types.

Acknowledgements

This work was financed by CONACyT (Mexican Science and Technology National Council) through 2015 CONACyT SEP-CB (Basic Science-Public Education Ministry) project fund 258553/CONACyT/CB-2015-2101. Thanks go to the Scientific Computing Laboratory at FCQ-UJED for computational resources. Thanks go to the Academic Group UJED-CA-129 for valuable discussions.

Conflict of interest

The authors declare no conflict of interest.

Author details

Diana Barraza-Jiménez[1], Sandra Iliana Torres-Herrera[2], Patricia Ponce Peña[1], Carlos Omar Ríos-Orozco[3], Adolfo Padilla Mendiola[1], Elva Marcela Coria Quiñones[3], Raúl Armando Olvera Corral[1], Sayda Dinorah Coria Quiñones[4] and Manuel Alberto Flores-Hidalgo[1*]

1 Laboratory for Scientific Computation, Faculty of Chemistry Science, Juarez University of Durango State, Durango, México

2 Faculty of Forestry Science, Juarez University of Durango State, Durango, Dgo., México

3 TecNM/Durango Institute of Technology, Durango, Dgo., Mexico

4 Cinvestav-IPN, Unidad Querétaro, Qro., Mexico

*Address all correspondence to: manuel.flores@ujed.mx

IntechOpen

References

[1] Frank A. Coutelieris, J. M. P. Q. Delgado. Transport Processes in Porous Media. Springer-Verlag Berlin Heidelberg 2012. DOI 10.1007/978-3-642-27910-2.

[2] Lage, J.L., Narasimhan, A.: Porous media enhanced forced convection fundamentals and applications. In: Vafai, K. (ed.) Handbook of Porous Media. Marcel Dekker, New York (2000).

[3] Ming-Hui Sun, Shao-Zhuan Huang, Li-Hua Chen, Yu Li, Xiao-Yu Yang, Zhong-Yong Yuan and Bao-Lian Su. Chem. Soc. Rev., 2016,45, 3479-3563.

[4] Arne Thomas. Nature Communications. (2020) 11:4985.

[5] Zaiku Xie, Bao-Lian Su. Front. Chem. Sci. Eng. 2020, 14(2): 123–126.

[6] Bao-Lian Su, Clément Sanchez, and Xiao-Yu Yang. Hierarchically Structured Porous Materials: From Nanoscience to Catalysis, Separation, Optics, Energy, and Life Science, 1st Edition. 2012 Wiley-VCH Verlag GmbH & Co. KGaA. 2012. Wiley-VCH Verlag GmbH & Co. KGaA.

[7] Jonah Erlebacher and Ram Seshadri. MRS Bulletin. 34. 2009. 561-570.

[8] Ruren Xu, Wenqin Pang, Jihong Yu, Qisheng Huo, Jiesheng Chen. Chemistry of Zeolites and Related Porous Materials: Synthesis and Structure. John Wiley & Sons (Asia). 2007. ISBN 978-0-470-82233-3.

[9] C. Feng, K.C. Khulbe, T. Matsuura, R. Farnood, A.F. Ismail. J. of Membrane Science and Research 1 (2015) 49-72.

[10] A. Maghfirah a, M.M. Ilmi a, A.T.N. Fajar b, G.T.M. Kadja. Materials Today Chemistry 17 (2020) 100348.

[11] K. Zhang, M.L. Ostraat. Catal. Today 264 (2016) 3e15, https://doi.org/10.1016/j.cattod.2015.08.012.

[12] Database of Zeolite Structures; http://www.iza-structure.org/databases/ (Accessed November 20, 2020).

[13] J.P. Ramirez, C.H. Christensen, K. Egeblad, C.H. Christensen, J.C. Groen. Chem. Soc. Rev. 37 (2008) 2530e2542, https://doi.org/10.1039/B809030K.

[14] B. Smit, T.L.M. Maesen. Nature 451 (2008) 671e678, https://doi.org/10.1038/nature06552.

[15] J. Jae, G.A. Tompsett, A.J. Foster, K. D. Hammond, S.M. Auerbach, R.F. Lobo, G.W. Huber. J. Catal. 279 (2011) 257e268, https://doi.org/10.1016/ j. jcat.2011.01.019.

[16] S. Teketel, L.F. Lundegaard, W. Skistad, S.M. Chavan, U. Olsbye, K.P. Lillerud, P. Beato, S. Svelle. J. Catal. 327 (2015) 22e32, https://doi.org/10.1016/j. jcat.2015.03.013.

[17] M.M. Recio, J.S. Gonzalez, P.M. Torres. Chem. Eng. J. 303 (2016) 22e30, https://doi.org/10.1016/j.ce j.2016.05.120.

[18] C. Song, Y. Chu, M. Wang, H. Shi, L. Zhao, X. Guo, W. Yang, J. Shen, N. Xue, L. Peng, W. Ding. J. Catal. 349 (2017) 163e174, https://doi.org/10.1016/ j.jcat.2016.12.024.

[19] M. Koehle, Z. Zhang, K.A. Goulas, S. Caratzoulas, G.D. Vlachos, R.F. Lobo. Appl. Catal. Gen. 564 (2018) 90e101, https://doi.org/10.1016/j.apcata .2018.06.005.

[20] X. Li, R. Xu, Q. Liu, M. Liang, J. Yang, S. Lu, G. Li, L. Lu, C. Ind. Crop. Prod. 141 (2019) 111759, https://doi.org/ 10.1016/j.indcrop.2019.111759.

[21] E.T.C. Vogt, B.M. Weckhuysen. Chem. Soc. Rev. 44 (2020) 7342e7370, https://doi.org/10.1039/C5CS00376H.

[22] Yi Li, Hongxiao Cao, and Jihong Yu. ACS Nano 2018, 12, 5, 4096–4104.

[23] Bereciartua, P. J.; Cantín, Á.; Corma, A.; JordÁ, J. L.; Palomino, M.; Rey, F.; Valencia, S.; Corcoran, E. W.; Kortunov, P.; Ravikovitch, P. I.; Burton, A.; Yoon, C.; Wang, Y.; Paur, C.; Guzman, J.; Bishop, A. R.; Casty, G. L. Science 2017, 358, 1068–1071.

[24] Jeon, M. Y.; Kim, D.; Kumar, P.; Lee, P. S.; Rangnekar, N.; Bai, P.; Shete, M.; Elyassi, B.; Lee, H. S.; Narasimharao, K.; Basahel, S. N.; Al-Thabaiti, S.; Xu, W.; Cho, H. J.; Fetisov, E. O.; Thyagarajan, R.; DeJaco, R. F.; Fan, W.; Mkhoyan, K. A.; Siepmann, J. I.; et al. Nature 2017, 543, 690–694.

[25] Jiao, F.; Li, J.; Pan, X.; Xiao, J.; Li, H.; Ma, H.; Wei, M.; Pan, Y.; Zhou, Z.; Li, M.; Miao, S.; Li, J.; Zhu, Y.; Xiao, D.; He, T.; Yang, J.; Qi, F.; Fu, Q.; Bao, X. Science 2016, 351, 1065–1068.

[26] Snyder, B. E. R.; Vanelderen, P.; Bols, M. L.; Hallaert, S. D.; Böttger, L. H.; Ungur, L.; Pierloot, K.; Schoonheydt, R. A.; Sels, B. F.; Solomon, E. I. Nature 2016, 536, 317–321.

[27] Shan, J.; Li, M.; Allard, L. F.; Lee, S.; Flytzani-Stephanopoulos, M. Nature 2017, 551, 605–608.

[28] Primo, A.; Garcia, H. Chem. Soc. Rev. 2014, 43, 7548–7561.

[29] Al-Khattaf, S.; Ali, S. A.; Aitani, A. M.; ŽilkovÁ, N.; Kubička, D.; Čejka, J. Catal. Rev.: Sci. Eng. 2014, 56, 333–402.

[30] Vogt, E. T. C.; Weckhuysen, B. M. Chem. Soc. Rev. 2015, 44, 7342–7370.

[31] Beale, A. M.; Gao, F.; Lezcano-Gonzalez, I.; Peden, C. H. F.; Szanyi, J. Chem. Soc. Rev. 2015, 44, 7371–7405.

[32] Li, Y.; Li, L.; Yu, J. Chem. 2017, 3, 928–949.

[33] Li, Y.; Yu, J. Chem. Rev. 2014, 114, 7268–7316.

[34] Neal S. Eash, Thomas J. Sauer, Deb O'Dell, Evah Odoi. Soil Science Simplified 6th Edition. 2016. John Wiley & Sons.

[35] Gregory, P. J. Eur. J. Soil Sci. (2006) 57:2–12.

[36] Garrison Sposito. The Chemistry of Soils. 2008. Oxford University Press, Inc.

[37] ANSYS Fluent® Academic Research CFD, Release 18.2, 19.1.

[38] ANSYS Fluent®. Customization Manual. USA: Published for licensees' users by ANSYS® INC; 2017

[39] G. K. Batchelor. An Introduction to Fluid Dynamics. Cambridge Univ. Press. Cambridge, England. 1967.

[40] PTC Creo Parametric, release 7.0. 2020. PTC Inc.

Chapter 6

External Flow Separation

Chandran Suren and Karthikeyan Natarajan

Abstract

The flow transit from laminar to turbulent over the surface due to adverse pressure gradient, that the region in between the laminar separation and turbulent reattachment is called Laminar separation bubble. It experiences on the many engineering devices as well as controls the aerodynamic and heat transfer characteristics. The way of transition formation differs based on geometry, flow configuration and method of transition initiations by a wide range of possible background disturbance as free stream turbulence, pressure gradient, acoustic noise, wall roughness and obstructions, periodic unsteady disturbance so on. This chapter discusses about the flow transition on airfoil and nozzle in general and focuses more on the transition process in the free shear layer of separation bubbles, free stream turbulence, and identification of separation point with the help of the CFD method.

Keywords: laminar separation bubble, turbulence intensity, turbulence separation, reattachment

1. Introduction

The transition from laminar to turbulent flow is extremely difficult and has eluded engineers, physicists, and mathematicians for more than a century, despite massive efforts. Given that transition may be observed in many engineering flows and has a significant impact on the aerodynamics and heat transfer properties of those flow systems, it has attracted a great deal of attention and research. Transition is so complicated because it can take many different paths depending on flow configuration and geometry, and the presence of many different flow disturbances, such as wall roughness or obstructions, free-stream turbulence, acoustic noise, pressure gradient, surface heating or cooling, suction or blowing of fluid from the wall, and so on, greatly influences the transition process.

It is commonly known that Osborne Reynolds was the first person to conduct systematic experimental studies of pipe flow transition in the late 19th century. William McFadden Orr and Arnold Sommerfeld pioneered the study of transition. They separately created a mechanism for explaining the start of turbulence, which was dubbed the Orr-Sommerfeld approach in their honor (more commonly known as linear stability theory). Active theoretical research of transition began in the early 20th century, following the Orr-Sommerfeld paradigm.

Numerical studies of transition, which included calculating a simplified, linearized version of the Navier–Stokes equations numerically, began long before computers were invented. Nonetheless, effective numerical studies of transition employing large-eddy simulation (LES) and direct numerical simulation (DNS) did not begin to emerge until the late 1980s, with significant development in the last two decades.

Transition has traditionally been divided into three broad types for wall-bounded flows.

1.1 Natural transition

When the free-stream turbulence intensity is less than 0.5% in an associated boundary layer, this transition happens. The transition process is initiated by two-dimensional (2D) instability waves known as Tollmien-Schlichting (TS) waves (primary instability), which are then followed by a three-dimensional (3D) instability (secondary instability), which leads to significant 3D flows with the formation of streamwise/spanwise vortices. The final phase of this transition process is known as the advance phase, which involves the collapse of these large-scale eddies into smaller flow structures and the creation of turbulent points that eventually merge into a turbulent boundary layer [1–6]. Natural transition is the most studied area compared to the other two categories of transition (a lot of work had already been done in the first half of the 20th century, mainly based on linear stability theory and some experiments) and therefore the transition process gets relatively much better understood. There are several stages involved in the transition process:

  i. Receptivity stage – how disturbances are projected into developing eigen modes, or how they enter or otherwise cause disturbances in a boundary layer, are all examples of responsiveness stage.

  ii. Primary instability – Small disturbances are amplified because of a primary instability (2D TS waves) in the flow, which is caused by the flow.

  iii. Secondary instability –usually, once the disturbance reaches a finite amplitude, it

  iv. Breakdown stage – Nonlinearities and perhaps greater instabilities cause the flow to excite a growing number of scales and frequencies. A growing number of scales and frequencies are induced in the flow by nonlinearities, with the possibility of increased instabilities.

1.2 Bypass transition

For boundary layers installed under sufficiently high flow disturbances, such as boundary layers on flat plates without a pressure gradient under turbulent free flow intensity greater than 1%, the transition occurs more rapidly and the 2D instability stage of the natural transition is skipped. Morkovin [7] coined the term "bypass transition" to describe this sort of transition. The skip transition turned into to start with visible as a mystery, because the turbulent factors have been created out of nowhere right away as compared to the natural transition. In the last two decades, numerical simulations with stability analysis (especially DNS - direct numerical simulation) have improved knowledge of bypass transition [8–16]. **Despite** the numerous proposed shunt transition mechanisms, a commonly regular description of shunt transition may be summarized as follows:

  i. At high levels of free-flow turbulence, low-frequency disturbances penetrating the laminar boundary layer may undergo algebraic growth (called transient growth or non-modal growth, with further reference to the fact that this modality is not predicted as from the automode of the linear theoretical solution based on the equation Orr Sommerfeld and Squire), which leads to the formation of longitudinal stripes. After P.S. Klebanoff, who was the first to analyses this phenomena and define it as a periodic thickening/thinning of the boundary layer [17, 18], these streaks are known as boundary layer streaks or Klebanoff distortions (or Klebanoff modes). They are disturbance zones similar to the forward and backward beams (high and low velocity lines) in the flow direction, which alternate in the wavelength direction with the wavelength of the order of boundary layer thickness.

ii. A laminar boundary layer with streaks is unstable, and the streaks expand in length and amplitude downstream. The transition is generally initiated close to the pinnacle of the boundary layer via few forms of inaction instability because of the interplay among low-speed tracks lifted from the close to wall area and excessive frequency disturbances within side the unfastened flow, that's strongly damped by laminar shear. (Known as shielding shear) and it consequently cannot penetrate the boundary layer.

iii. After an initial burst of turbulent activity, a boundary layer is formed when the early bursts of turbulent activity combine.

1.3 Separated-flow transition

When the laminar boundary layer separates or when the laminar flow separates on the beveled/sharp/rounded front edge of the flat plate, transitions can occur in the separate flow-free shear layer. This is referred to as a separated-flow transition (or called separated boundary layer transition). It's worth mentioning that transition has been classified into four kinds in some literature, the fourth of which is termed wake-induced transition [6, 19–22], since in turbomachinery flows, impinging wakes from the previous blade rows greatly influence the changeover process.

The use of a single category (separate flow transition) to describe the transition in a separate laminar shear layer is too general or too vague. Walker [23] suggested in the early 1990s that, like the bypass transition in a connected boundary layer, the "bypass transition" might also occur in separated shear layers. Despite this, research on the issue of "bypass transition in separated shear flows" has been quite sparse. Furthermore, the separation may be brought on in numerous ways, for example, separation of the boundary layer on a flat plate because of and a mainstream gradient; geometrically brought on separation, or even in a few instances the separated streams re-connect to the floor to form separation bubbles whilst in different instances they by no means re-connect.

Section 2 of this study will explore the transition process in separation bubbles created by an unfavorable pressure gradient. Section 3 will focus on the influence of free turbulence levels on the transition process in geometrically induced separation bubbles where the separation point is a very short distance for the development of an attached boundary layer or no boundary layer development for the point of separation at all.

2. Transition in separation bubbles induced by an adverse pressure gradient

Adverse pressure gradient may cause the connected laminar boundary layer to split, resulting in a free turbulence turns in to forming position and transit into a stormy layer. This layer can be found in gas turbine blades at internal flows and at external flow it was found in aircraft wings and wind turbine blades. The stream turbulent intensity of compressor/turbine regions was 20% which was greater compared to turbo machinery flows as nearly 5–10% and for wind turbine blade it was lesser than 5% [24].

2.1 Transition process under low free-stream turbulence

Experimental and numerical studies clearly describe about the free stream turbulent intensity in which the transition process occurs at this condition is due to without viscosity, the state which is likely to change and Kelvin-Helmholtz (KH) instability [25–32]. On comparing the TS wave are often lower than the rate of the upstream 2D instability waves propagate downstream. Additionally, lower stream, and tangential uncertainty process linked along with the deformation orderly 2D directed whirlwind as the KH rolls is responsible for the development of 3D movements. Because of the deformed KH rolls, a stream-wise vorticity develops in the stream. At the macro level, the large-scale coherent structures disintegrate into smaller size structures that tend to disrupt. This brings about turbulence in the mean reattachment point area. The TS uncertainty may be still there and interact with the KH uncertainty during the separated boundary layer transition process, and the TS uncertainty would plays an important role in the failure of the turbulent in some cases, despite the fact that KH uncertainty usually plays a presiding role in the procedure [33–37].

In the squat level of the free flow stormy, we exit relatively unlike a fixed physical phenomenon with the first step of the procedure described above - the primary stage of uncertainty is well understood - has the perception of the transitional procedure in a detachment illusion. In the attached physical phenomenon passage, like K-type secondary instability, H-type secondary instability, or O-type secondary instability has reasonably well-understood secondary instabilities, whereas for detached flow transition, the existing knowledge of secondary instabilities is vague and numerical studies performed on imposed and freewill flat separation bubbles on a pinion occurrence found that the two possible secondary uncertainties were vigorous: elliptical uncertainty in curt region of the moisture flow and hyperbolic uncertainty in an articulate region of the moisture flow [27, 38–42].

2.2 Transition process under elevated free-stream turbulence

This work shows that turbulent intensity has a major impact on bubble entrainment that is increasing mean shear layer, reducing mean bubble length, making turbulent flow motion as early [24, 43–47].

Haggmark [48] done an experimental investigation into a flat plate under a grid, the separation bubble developing generated a turbulent intensity of 1.5 percent, and boundary layer streaks with less frequency in the outer layer and free shear layer which also causes high amplitude. Additionally, the two-dimensional waves created in the experiment by the flow visualization were not found. McAuliffe and Yaras [28] determined the effects of pressure gradient on separation bubbles created in a level plate with low pressure gradient at 0.1% and high-pressure gradient at 1.45% free shear turbulence levels by introducing DNS. Numerical Simulations were conducted with CFD code: ANSYS CFX. For spatial and temporal discretization, the second-order central and Euler backward differencing scheme was used. Primary finding is that at the low free-stream shear turbulence level, the acceptance of the laminar separated free shear layer happens through KH instability mechanism by minor disturbances but in high free-stream turbulence level, the separation of upstream occurs due to the rolling up of free shear turbulence layer because of boundary layer streaks. When the separated free shear layer interacts with streaks through a localized secondary instability, turbulent spots are generated.

Balzer and Fasel [24] investigated the boundary layer separation caused due to free stream turbulence level by using DNS. The non-permanent derivatives were discretized using an explicit, Runge–Kutta method, and the spatial derivatives using compact differences. The turbulent intensity of various three cases was studied: 0.05%, 0.5%, and 2.5%. These case studies show that in the laminar boundary layer, stretched streaks appear at even the lowest free-stream turbulent intensity of 0.05% as shown in **Figure 1**. In this figure, the regions in which the Klebanoff mode is associated with acceleration (u040) and deceleration (u0o0). For 2.5% free-stream turbulent intensity it was found that the amplitude of these streaks increased significantly when the level of open stream turbulent intensity increased as shown in **Figure 2**.

While at 2.5 percent free-stream turbulent intensity the streaks are closely linked to the turbulent flow structures in the area 12oxo13, at 0.05 percent free-stream turbulent intensity the Klebanoff modes are weak and not directly connected. Additionally, the researchers discovered that separation continued to occur even in the case with the highest free-stream turbulence intensity (2.5 percent), and upstream of the separating bubble had no turbulence areas The weird review expose such that the occasion without free stray (top of **Figure 3**), a distinct pit is clearly observed, linked to the invisible shear layer uncertainty (KH uncertainty). In addition to the higher free stream turbulence, this distinctive peak is still observed at a similar value as shown in **Figure 3** (bottom). This confirms that, as demonstrated by the linear stability analysis, even in the event of the highest free-stream turbulent intensity (2.5%), the KH instability mechanism is still present. Therefore, they concluded that either the foremost curtail layer uncertainty and the increased 3D disburse level, peculiarly in the spurt wise trait occurred by free stream turbulence, were responsible for the turbulence flow. Results from the new numerical review study conducted by Li and Yang [49] of a separate adaptation to a border layer on a board like, with a turbulence nature vigor of 3%. Similarly, to the findings made previously by Li and Yang [49] that barrier layer trait (Klebanoff distortions or modes) was initiated crucial of the disunion. But even so, visualization reveals that the KH uncertainty is active, proving that there is also distortion in the KH flow. In a certain tranche of the KH roll, it can merge along the traits together with the

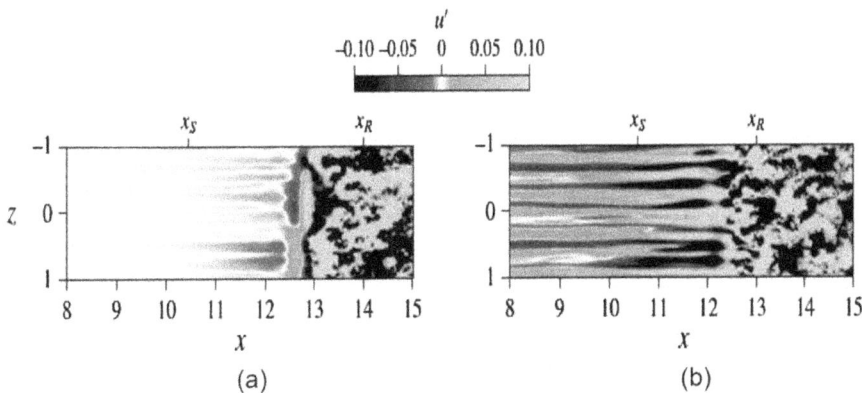

Figure 1.
Gray scale contours of the stream-wise velocity variations for free-stream turbulence intensities of 0.5 and 2.5 percent in an x-z plane at y = 1(x) at y = 1(x) [24].

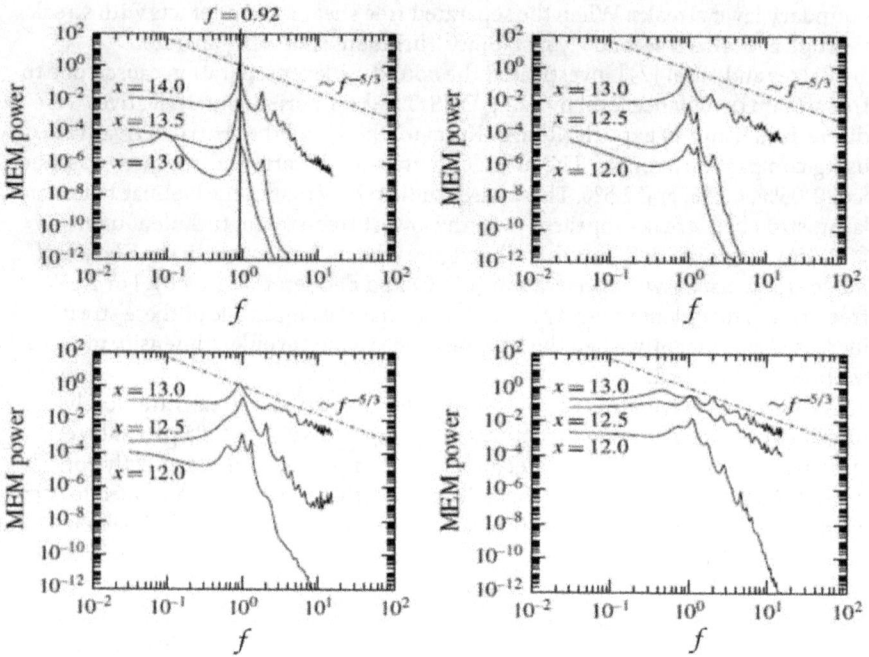

Figure 2.
Maximum entropy spectral energy at various free-stream turbulence intensities at 0; 0.05; 0.5; 2.5 ms [24].

confused 3D structure swiftly ensuing. The existence of the KH uncertainty was established by their further stability investigation.

Istvan and Yarusevych [47] carried out an experiment on the adaptation from stratified flow in the bubble which is produced above the siphon side of a NACA 0018 pinion at 2 different Reynolds numbers (80,000, 125,000). To monitor drizzle growth on two of the **spurts** wise and spread the axes while investigating free stream turbulence vigor levels (0.06%, 0.32%, 0.51%, and 1.99%). The findings of them, indicate that the curtail layer is only rolling upward in vortexes to the middle transition point, which led to greater downstream vortex shedding, according to **Figure 4**. The site of curtail layer furl is pushed crucial as free stream turbulence vigor **increases but** curtail layer furl/swirl discard is still lucid detected as illustrated in **Figure 4(d)**, compared **to the** observed in bottommost free-stream turbulence adaptation scenarios. Here, it suggests that the adiabatic curtail layer uncertainty (KH uncertainty) is **still** active beneath conditions of 1.99% free-stream turbulence intensity. However, like the appropriate orthogonal decomposition (PED) review, the wise streams spread from the partition layer crucial from the bubble is a significant role in the adaptation procedure for the higher cases of free-stream turbulence adaptation **since** the streams are **provided** to the spirit of speed flushes rather than the Spanish rollers.

Zaki et al. [50] conducted a thorough investigation on the impact of free-stream turbulence on adaptation in a compressor cascade. Five instances were evaluated, one without turbulence and four with varying turbulence adaptation at the creek: 3.25%, 6.5%, 8.0%, and 10%. The adaptation procedure observed on the strain surface is much differ from the observed on the siphon surface, since disunion happens barely in the absence of turbulence at the creek, whereas flow stays attached in all other instances. **Under** free-stream turbulence levels, the partition layer converts to turbulence crucial of the stratified separation site, guaranteeing that drizzle

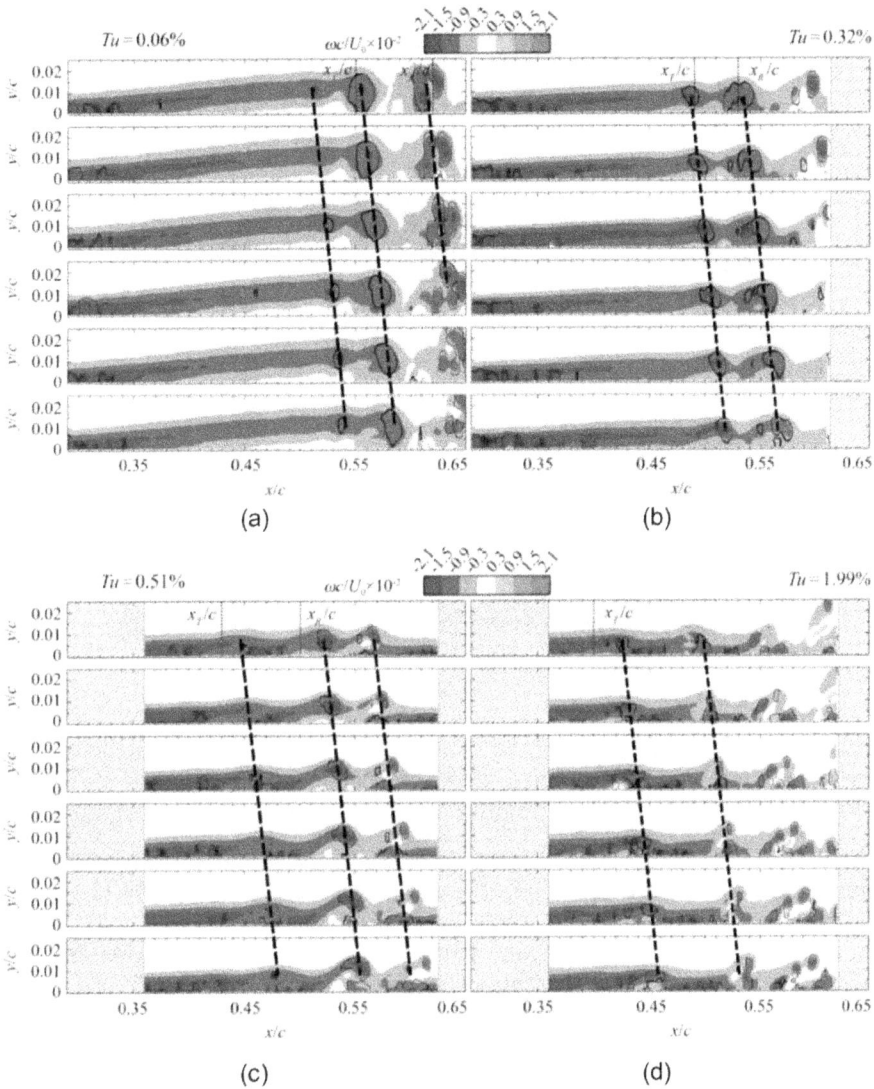

Figure 3.
Spanwise vorticity contours, dotted lines trace estimated centres of spanwise vorticity, and images spaced by 0.33 ms [24].

attachment continues. In additional investigations, wake-initiated adaptation occurred previous the partition layer can divide [51, 52]. According to the findings of Zaki et al. [50], traits were observed and magnified as far crucial as 3.25% of the entrance turbulence strength (decompose to about 2.5% at the blade dominant lead). They discovered, the breakup of those trait did not go after the usual detour process, as an internal uncertainty comparable to that exhibited in classical TS waves' subsidiary uncertainty was identified. However, they further revealed that the transition through the secondary instability of the streaks at greater inlet turbulence intensities circumvent was dominating. However, laminar separation **continues** the siphon surface and is regulated by partition layer traits that emerge crucial of the disunion position when the input turbulence adaptation is 3.25%. **After** the KH roll formation, they tend to become unstable and breakdown to turbulence.

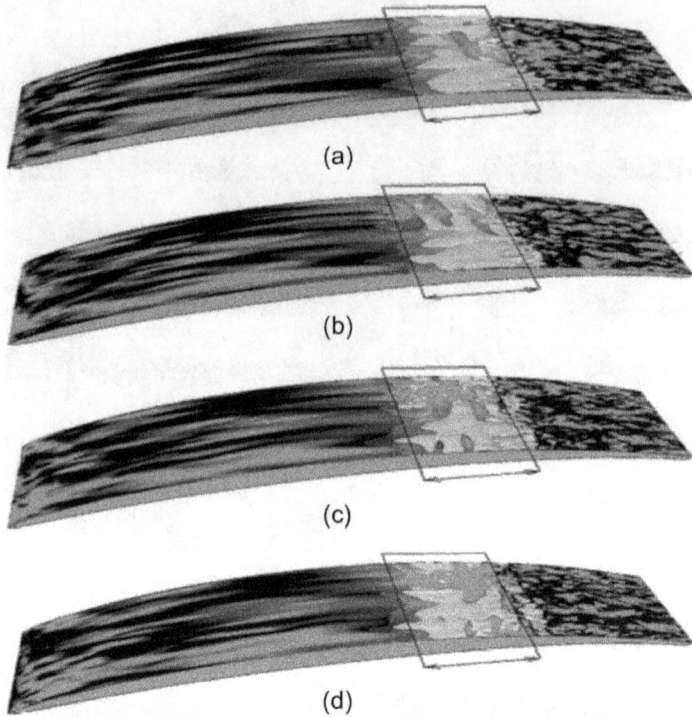

Figure 4.
Four tangential velocity variations contours superimposed on the instantaneous separation surface (white area). Dark lines represent the mean separation length [47].

"Also, Li and Yang [49] have shown that the KH rolls are not **2D but** have extremely deformed fingers instead. The average results at the next greater level for turbulence inflow strength of 6.5% (declining to roughly 4% at the border of the.

Blade) suggest stratified disunion and the successive turbulence reassembly. They pinned that, however, that this was deceptive because the rapid drizzle meadow revealed the creation of turbulence smudge in few places where the partition layer remains fixed, as illustrated in **Figure 5**. Drizzle disunion occurs over the full span in the first two occurrence, as shown in **Figure 5(a)** and **(b)**, however in the other two instances, as shown in **Figure 5(c)** and **(d)**, drizzle disunion occurs just at a specific spread region and flow remains attached across the other region.

When the inlet turbulent intensity was further increased to 8% and 10% (decaying to 5.5% and 6.3% at the blade leading edge), they showed that the partition layer adaptation was associated with mean flow curvature and the adaptation was driven primarily by the detour procedure of an attached partition layer.

Simoni et al. [46] conducted experimental research employing time-resolved PIV instrumentation to evaluate the impact of free-stream turbulence adaptation magnitude on the composition and energetic features of a stratified disunion bubble at 3 different Reynolds numbers. Owing to unfavorable constraint ramp characteristic of ultra-high-lift piston blade outline, the bubble developed on a flat plate. They have been measured by three Reynolds (40,000; 75,000; 90,000) and 3 freely flow turbulence levels on the plate's superior contour in the midspan (0.65%, 1.2% and 2.87%).The findings demonstrate that the disunion bubble shrinks as the Reynolds number and free-stream turbulence adaptation increase, with the separation bubble shrinking the most at the greatest Reynolds number (90,000)

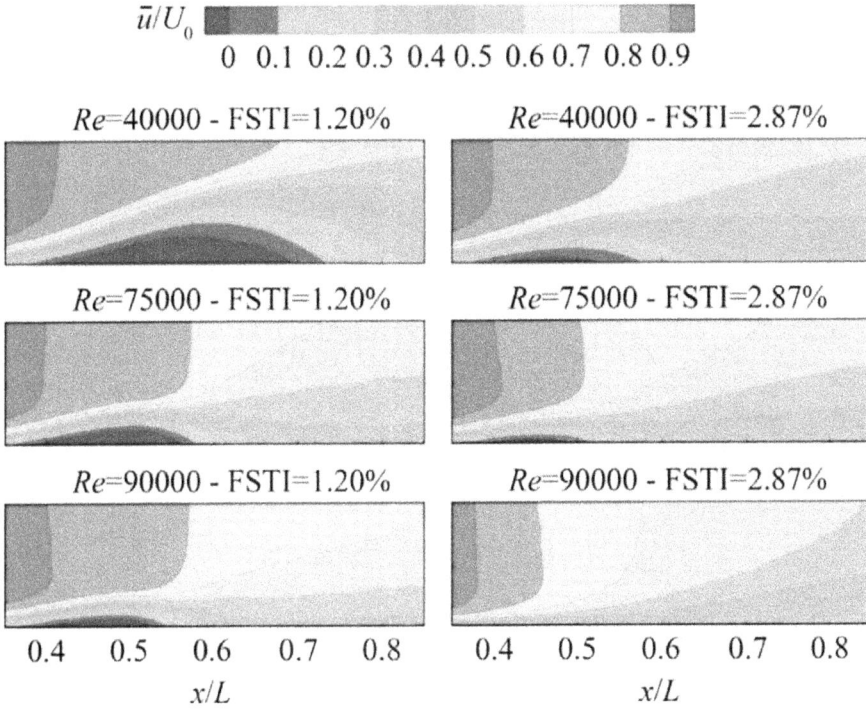

Figure 5.
Mean normalized streamwise velocity contours for two free-stream turbulence levels (1.2%, 2.87%) [50].

and free-stream turbulence adaptation (2.87 percent) The imply spurt wise fleetness lineation depicted in **Figure 6** did not reveal the presence of the bubble. For all save the lofty Reynolds number (90,000) and free-stream turbulence adaptation, the vortex shedding frequency observed was owing to the adiabatic curtail layer

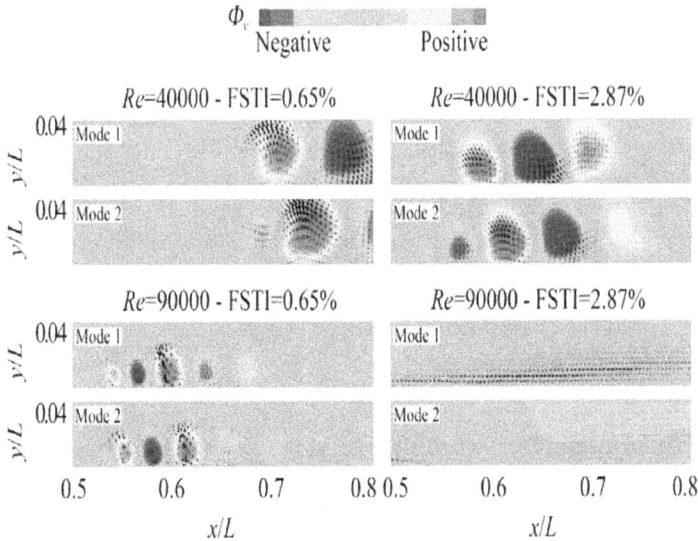

Figure 6.
POD modes and iso-contour lines [46].

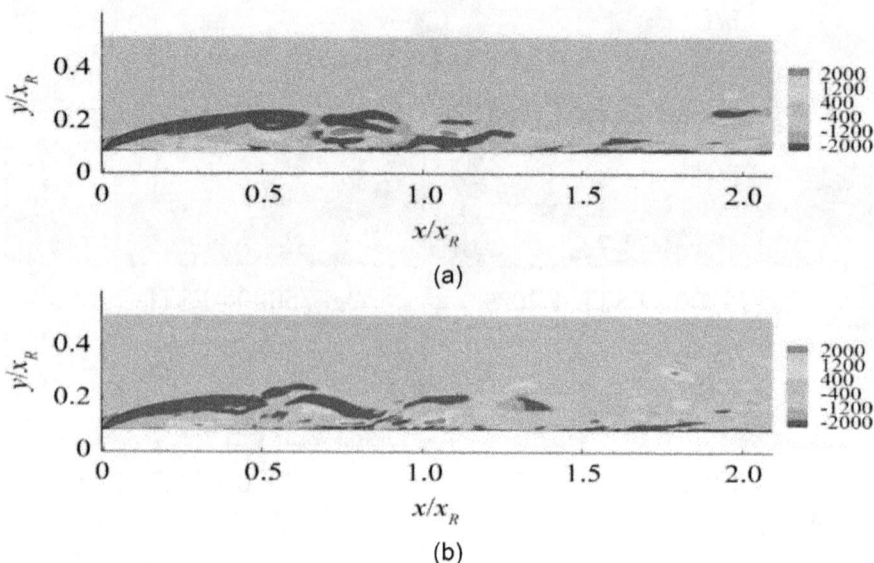

Figure 7.
(a) No free-stream turbulence case and (b) 2 percent free-stream turbulence intensity scenario [46].

uncertainty (KH uncertainty) (2.87%). Proper Orthogonal Decomposition (POD) study confirmed the fact that vortex shedding did not occur at the maximum Reynolds number and free-stream turbulence shedding as shown in **Figure 7**. In this figure it is evident that POD modes are representing the turbulence shedding process in all circumstances, save from the largest number of Reynolds and free-strip turbulence shedding case. **Figure 6** also confirms that when the free-stream turbulence intensity is increased to 2.87% at the Reynolds number of 90,000, the separation bubble is completely removed.

Previous investigations have showed that transition occurs more quickly when free-stream turbulence is elevated, in addition with a type of "detour adaptation" that considered in various research as a possible explanation. However, "detour adaptation" here refers to disturbances entering the disunion curtail layer and detour the curtail layer furl [28]. Whereas the major direct uncertainty procedure, TS Waves, is detoured through a detour adaptation in an associated partition layer. There is currently no indication that the direct curtail layer uncertainty procedure, the KH uncertainty, is detoured in the disunion partition layer adaptation under elevated free-stream turbulence. Several research have confirmed the presence of the KH instability up to 3% free-stream turbulent intensity, and a few investigations have showed that separation is suppressed at much greater free-stream turbulent intensities and/or at higher Reynolds numbers. Therefore, saying that KH uncertainty is detoured because a disunion curtail layer never exist longer is not a fair representation of the situation.

3. Transition in separation bubbles induced geometrically

The indicated part concentrates on the detached bubble adaptation geometrically. Detached bubbles are distinguished by the short distance that separates the separation point in the initial stage of adhere frontier layer, e.g., detached bubbles on the board like with a blunt/rounded leading edge [53–56]. The adaptation could occur only in the detached shear layer because of this. In several studies [57–65], it became clear that the adaption procedure is launched by the mechanism of KH uncertainty identical as in the situations detailed in the previous portion, under the low freely-stream turbulence. The transition process, on the other hand, can be considerably unlike under raised free-stream turbulence because at that point essentially nay connected extremities layer building before detachment.

Due to the lack of research, it is unclear how exactly the separation process takes place in separation bubbles. Experiments have been conducted on interim disunion bubble on a board like with a semicircular indigenous under various flow position and free stream turbulence levels, dubbed the T3L test case by the ERCOFTAC Special Interest Group on Transition [53, 66], but regrettably few complete quantitative effects on this test case have been performed.

The large eddy simulation (LES) from Yang and Abdallah [67, 68] was numerically used to study transitional separate–attached flow across flat plate with a blunt ledge of less than 2% of freestream turbulence intensity. The control equations were disconnected on a stumbled grid, and the sub grid pressure were resembled by an effective sub grid scale model. The crystal clear second order Adams–Bashforth temporal strategy was utilized for the mortal discretization, and the secondary order second order central differencing spatial scheme was used for temporal discretization. The research revealed that when free-stream turbulence is present, the mean bubble length is lesser by around 14% [61]. When no free-stream turbulence was present, the layer separating the flow parted earlier, as can be seen in **Figure 8**, where instantaneous span

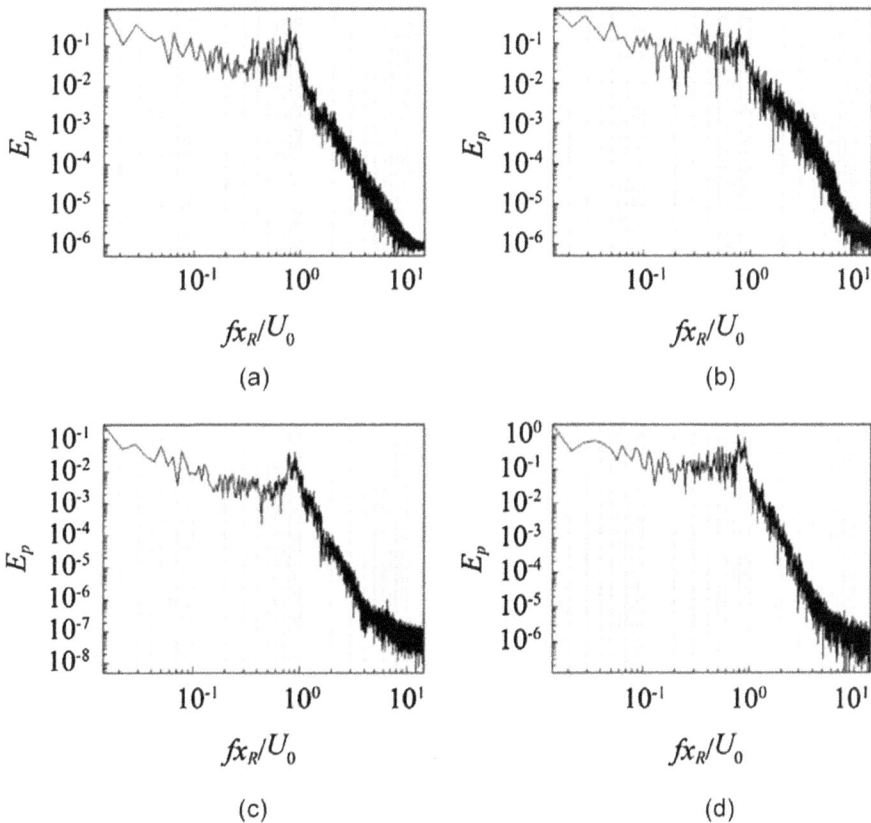

Figure 8.
The pressure spectra at $x/x_R = 0.75$ and at (a) $y/x_R = 0.01$, (b) $y/x_R = 0.05$, (c) $y/x_R = 0.13$, and (d) $y/x_R = 0.2$ [67].

wise vorticity is shown for the 2% free-stream turbulence case and a non-turbulence free-spurt scenario. Nonetheless, their flow visualization revealed the presence of the KH rolls, indicating the KH uncertainty. This was established by the presence of the aspect of peak in the spectrum depicted in **Figure 9**, which identifies the feature of the KH value [61]. **Figure 10** depicts the isometric view of instantaneous spanwise vorticity of low and high stream free turbulence.

Langari and Yang [69] conducted a comprehensive LES analysis on a flat, semi-circular edge transitional boundary layer, under two level turbulence-free streams (0.2% and 5.6% above the leading edge). A separate domestic restricted volume LES code was used to simulate the simulations. Pressure–velocity decoupling was eliminated using Rhie-Chow pressure smoothing. Spatial discretization was accomplished using a second order central differencing technique, while temporal discretization was accomplished using a single stage backwards Euler approach. An effective secondary network model was familiar with the secondary network pressure. They showed that the unsecured deformation layer created within the disunion bubble viz. the Kelvin Helmholtz uncertainty mechanism for the low free-stream turbulence, consistent with numerous prior research stated above, is invisibly unstable.

It is a tiny boundary layer that is susceptible to free-stream turbulence disturbances and generates small amounts of turbulent kinetic energy. The uncertainty

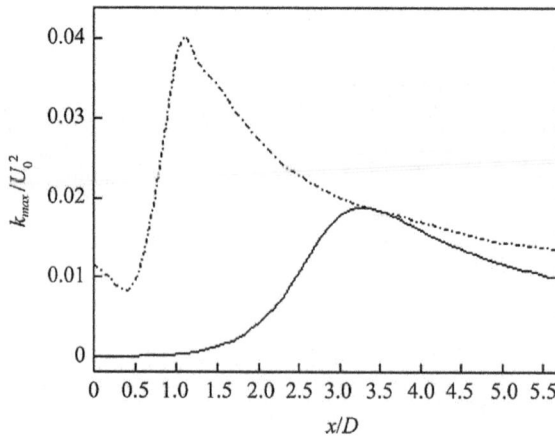

Figure 9.
Low free-stream turbulence scenario (solid line) and raised free-stream turbulence case (dotted line) (dashed line) [68].

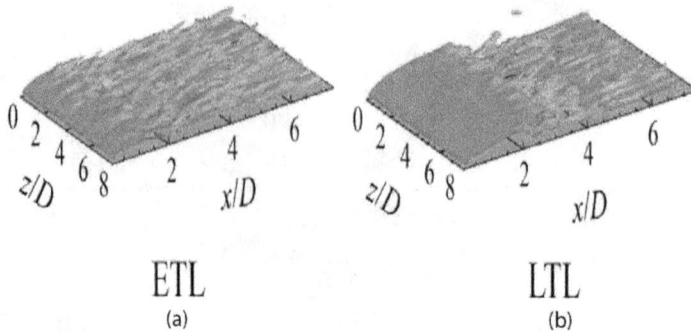

Figure 10.
In the situation of high free-stream turbulence and low free-stream turbulence, isometric view of instantaneous spanwise vorticity [69].

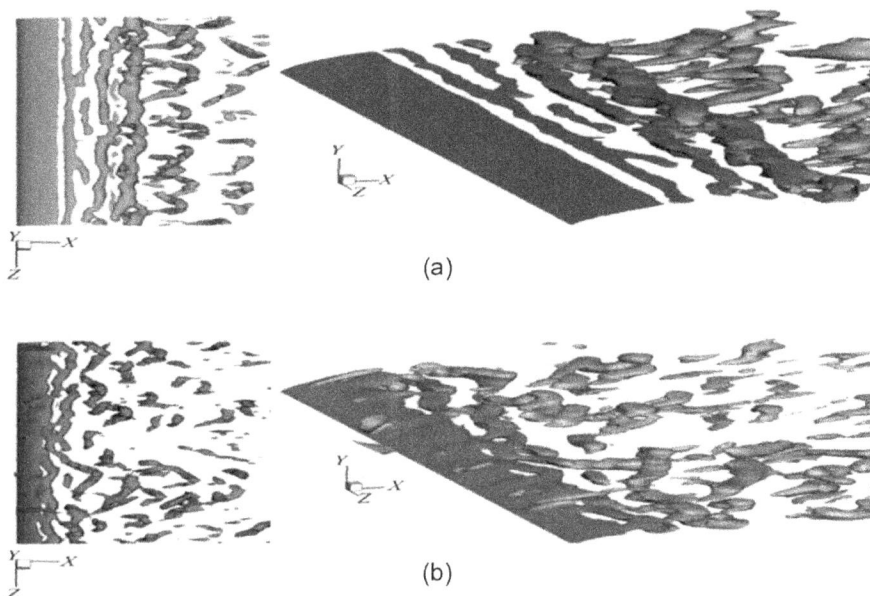

Figure 11.
Perspective views of the Q-criterion is surfaces: Examples of low free-stream turbulence and high free-stream turbulence are shown [69].

reviewed that the norm of the occurrence of the KH uncertainty is no longer appeased which forcibly suggests that the KH uncertainty has been detoured. This is farther proven by **Figure 11** theirs flux displays, which for the high level of free-stream turbulence (5, 6%), the early stage of the transition process differs greatly from the low free stream turbulence case (0.2%). This phenomenon, which is difficult to detect in the disunion curtail layer, makes the creation of 3D structures possible, and 3D structures arise in the separated shear layer much earlier in the formation of the bubble's initial turbulence than in the span wise oriented 2D KH rolls scenario. They conclude that "detoured adaptation" occurs as the KH uncertainty stage is detoured when TS intensity exceeds 5.6%.

The actuator creates a micro-jet that is pointed in the general direction of the free stream and is aligned at an angle to the surface. The jet's velocity changes on a regular basis in response to changes in the periodic excitation voltage. Periodic voltage signals at the required carrier frequency or pulsed actuation may be used to excite the actuator in this way, introducing periodic disturbances at a relevant hydrodynamic frequency into the flow. These two methods are referred to as continuous and pulsed mode, respectively [70].

An applied voltage signal range of 2 to 5 kV peak-to-peak was used to investigate the influence of excitation amplitude. It was determined that for each modulation frequency, the duty cycle was modified to generate the same momentum input every pulse regardless of the voltage amplitude. Furthermore, it is well known that the amplitude of the DBD plasma actuation is nonlinear to the applied voltage [71].

It can be seen that rms streamwise velocity variations are increasing in the separated shear layer and the reverse flow zone near to the wall. Aft in the bubble's streamwise shifting velocity profiles, distinct triple peaks develop [72].

The impacts of measurement noise in mean velocity fields are well-known to have an impact on LST estimates. Stability calculations propagate the experimental error to estimate the uncertainty in LST forecasts [73].

The flow becomes more steady and the amplitude of subsequent disturbances decreases as the bubble collapses. In this way, owing to the lower amplification rates and lesser momentum entrainment experienced by successive shocks, the smallest bubble state cannot be maintained. Because of this, the bubble grows in the second half of the transient to reach a quasi-steady state. To put it another way, when excitation is withdrawn from a bubble, it's more stable at its commencement of transitory than when it forms without it. By eliminating the excitation, the higher-amplitude harmonic disturbances are replaced by a larger spectrum of lower-amplitude natural perturbations, resulting in substantially smaller velocity variations in the bubble's aft part. The same mean bubble topology can no longer be maintained due to the decreased momentum entrainment, and the bubble essentially explodes, i.e., the size of the bubble quickly rises [74].

As a result, shear layer disturbances are more closely linked to the shedding frequency due to the repeated passing of vortices over the trailing edge, resulting in a larger upstream feedback loop. In comparison to the effects of manufactured disruptions, tone emissions have a little effect on the features of bubbles that form spontaneously [75].

4. Conclusion

This work gives an ideal review of the adaptation between the walls that have been categorized into three main groups: natural adaptation, bypass adaptation, separated-flow adaptation. Our present perception of those three major adaptation groups is summarized, with detached glide (or detached boundary layer) adaptation being the tiniest grasp in collation to natural adaptation and bypass adaptation in linked boundary layers.

This paper classifies two different groups of separation-flow transitions: those resulting from an adverse pressure gradient and those due to geometric separations. A boundary layer is present over a certain distance on the first sub-group of aircraft, but there is virtually no distance where one can be found for the secondary sub-group. It has been shown that the dominant main instability is KH uncertainty for the initial sort under truncated flow turbulent, but also TS uncertainty that interacts with KH uncertainty can be present. Some scientist used the parlance as "detour adaptation "to represent the adaptation process under elevated free-run turbulence but 'detour 'now does not mean that the KH stage is detoured because the KH level quiet exists below free run turbulent intensities of up to 3%. Even so the study has demonstrated increasing the free flow turbulent vigor to about 5.5%. Reduces the separations at the suction surface, with the separation at the suction surface with the separation influenced by the detour mechanism of adhere of the physical phenomenon. The indicated physical phenomenon is beard by several additional inspections that confirmed that the dissociation is in effect eliminated underneath adequately giant free stream turbulence. In adhere of the physical phenomenon and nay in an unoccupied shear layer, the usual detour occurs. In other words, it no longer means what it once did to say that KH uncertainty is detoured because the shear layer is now separated.

Many observations clearly reveal that the KH uncertainty is the superior technique in the secondary sub cluster under conditions of shallow free stream turbulence. Additionally, the evidence that compels the KH uncertainty phase surely detoured in the appearance of sufficiently strong free stream oil flow. In case of the secondary subgroup the detour adaptation occurs while unchained flow of turbulence potency is greater that is essentially distinct from the initial subgroup as the separation is eliminated in sufficient freight flow turbulence intensity.

Author details

Chandran Suren[1] and Karthikeyan Natarajan[2*]

1 Cape Institute of Technology, Tirunelveli, India

2 Department of Mechanical Engineering, National Engineering College, Kovilpatti, Tamilnadu, India

*Address all correspondence to: rsskarthikeyan@gmail.com

IntechOpen

References

[1] Mayle RE. The 1991 IGTI scholar lecture: The role of laminar-turbulent transition in gas turbine engines. Journal of Turbomachinery. 1991;**113**(4):509-536. DOI: 10.1115/1.2929110

[2] Comte Pierre FD, Lesieur M. Large-eddy simulation of transition to turbulence in a boundary layer developing spatially over a flat plate. Journal of Fluid Mechanics. 1996;**326**:1-36. Available from: https://www.cambridge.org/core/article/largeeddy-simulation-of-transition-to-turbulence-in-a-boundary-layer-developing-spatially-over-a-flat-plate/C277DE968A1FD929D3CB05FDBC434AAD

[3] Langari M, Yang Z. On transition process in separated-reattached flows. Adv. Appl. Fluid Mech. 2010;**8**:157-181

[4] Sayadi T, Moin P. Large eddy simulation of controlled transition to turbulence. Physics of Fluids. 2012;**24**(11):114103

[5] Schlichting H, Gersten K. Boundary-Layer Theory. Berlin, Heidelberg: Springer Berlin Heidelberg; 2017

[6] Dick E, Kubacki S. Transition models for turbomachinery boundary layer flows: A review. International Journal of Turbomachinery, Propulsion and Power. 2017;**2**(2):4

[7] Morkovin MV. On the many faces of transition. In: Proceedings of the Symposium on Viscous Drag Reduction. New York Dallas, USA: Springer Science+Business Media; 1968. pp. 1-31

[8] Voke PR, Yang Z. Numerical study of bypass transition. Physics of Fluids. 1995;**7**(9):2256-2264. DOI: 10.1063/1.868473

[9] Andersson P, Brandt L, Bottaro A, Dans H. On the breakdown of boundary layer streaks. Journal of Fluid Mechanics. 2001;**428**:29-60. Available from: https://www.cambridge.org/core/article/on-the-breakdown-of-boundary-layer-streaks/EB765552BE54B512449AD80076CD6EF8

[10] Jacobs RG, Durbin PA. Simulations of bypass transition. Journal of Fluid Mechanics. 2001;**428**:185-212. Available from: https://www.cambridge.org/core/article/simulations-of-bypass-transition/9AF1F9F2E9F9034A00A94287B1E59A01

[11] Zaki TA, Durbin PA. Mode interaction and the bypass route to transition. Journal of Fluid Mechanics. 2005;**531**:85-111

[12] Durbin P, Wu X. Transition beneath vortical disturbances. Annual Review of Fluid Mechanics. 2006;**39**(1):107-128. DOI: 10.1146/annurev.fluid.39.050905.110135

[13] Schlatter P, Brandt L, de Lange HC, Henningson DS. On streak breakdown in bypass transition. Physics of Fluids. 2008;**20**(10):101505. DOI: 10.1063/1.3005836

[14] Wang JJ, Pan C, Zhang PF. On the instability and reproduction mechanism of a laminar streak. Journal of Turbulence. 2009;**10**:N26. DOI: 10.1080/14685240902906127

[15] Hack MJP, Zaki TA. Streak instabilities in boundary layers beneath free-stream turbulence. Journal of Fluid Mechanics. 2014;**741**:280-315. Available from: https://www.cambridge.org/core/article/streak-instabilities-in-boundary-layers-beneath-freestream-turbulence/208709417DFB11CA962D15D9684EAB0B

[16] Xu Z, Zhao Q, Lin Q, Xu J. Large eddy simulation on the effect of free-stream turbulence on bypass transition. International Journal of

Heat and Fluid Flow. 2015;**54**:131-142. Available from: https://www.sciencedirect.com/science/article/pii/S0142727X15000569

[17] Klebanoff TK, D. United States. National Bureau of Standards. PS. Evolution of Amplified Waves Leading to Transition in a Boundary Layer with Zero Pressure Gradient. Washington [D.C.]: National Aeronautics and Space Administration; 1959

[18] Klebanoff PS. Effect of free-stream turbulence on a laminar boundary layer. In: Bulletin of the American Physical Society. 1305 Walt Whitman RD, STE 300, Melville, NY 11747-4501 USA: Amer Inst Physics; 1971. p. 1323-+

[19] Orth U. Unsteady boundary-layer transition in flow periodically disturbed by wakes. Journal of Turbomachinery. 1993;**115**(4):707-713. DOI: 10.1115/1.2929306

[20] Schobeiri MT, Read K, Lewalle J. Effect of unsteady wake passing frequency on boundary layer transition, experimental investigation, and wavelet analysis. Journal of Fluids Engineering. 2003;**125**(2):251-266. DOI: 10.1115/1.1537253

[21] Wissink JG, Zaki TA, Rodi W, Durbin PA. The effect of wake turbulence intensity on transition in a compressor Cascade. Flow, Turbulence and Combustion. 2014;**93**(4):555-576. DOI: 10.1007/s10494-014-9559-z

[22] Wu X, Jacobs RG, Hunt JCR, Durbin PA. Simulation of boundary layer transition induced by periodically passing wakes. Journal of Fluid Mechanics. 1999;**398**:109-153

[23] Walker GJ. The role of laminar-turbulent transition in gas turbine engines: A discussion. Journal of Turbomachinery. 1993;**115**(2):207-216. DOI: 10.1115/1.2929223

[24] Balzer W, Fasel HF. Numerical investigation of the role of free-stream turbulence in boundary-layer separation. Journal of Fluid Mechanics. 2016;**801**:289-321. Available from: https://www.cambridge.org/core/article/numerical-investigation-of-the-role-of-freestream-turbulence-in-boundarylayer-separation/7196ED62868D54E15E0D56EA0939DFA0

[25] Spalart PR, Strelets MK. Mechanisms of transition and heat transfer in a separation bubble. Journal of Fluid Mechanics. 2000;**403**:329-349

[26] Burgmann S, Dannemann J, Schröder W. Time-resolved and volumetric PIV measurements of a transitional separation bubble on an SD7003 airfoil. Experiments in Fluids. 2008;**44**(4):609-622. DOI: 10.1007/s00348-007-0421-0

[27] McAuliffe BR, Yaras MI. Passive manipulation of separation-bubble transition using surface modifications. Journal of Fluids Engineering. 2009;**131**(2):021201

[28] McAuliffe BR, Yaras MI. Transition mechanisms in separation bubbles under low- and elevated-freestream turbulence. Journal of Turbomachinery. 2009;**132**(1):011004. DOI: 10.1115/1.2812949

[29] Satta F, Simoni D, Ubaldi M, Zunino P, Bertini F. Experimental investigation of separation and transition processes on a high-lift low-pressure turbine profile under steady and unsteady inflow at low Reynolds number. Journal of Thermal Science. 2010;**19**(1):26-33. DOI: 10.1007/s11630-010-0026-4

[30] Dähnert J, Lyko C, Peitsch D. Transition mechanisms in laminar separated flow under simulated low pressure turbine aerofoil conditions. Journal of Turbomachinery. 2013;**135**(1):011007

[31] Yang Z. Numerical study of instabilities in separated-reattached flows. International Journal of Computational Methods and Experimental Measurements. 2013;**1**(2):116-131

[32] Serna J, Lázaro BJ. On the laminar region and the initial stages of transition in transitional separation bubbles. European Journal of Mechanics - B/Fluids. 2015;**49**:171-183. Available from: https://www.sciencedirect.com/science/article/pii/S0997754614001368

[33] Lang M, Rist U, Wagner S. Investigations on controlled transition development in a laminar separation bubble by means of LDA and PIV. Experiments in Fluids. 2004;**36**(1):43-52. DOI: 10.1007/s00348-003-0625-x

[34] Volino RJ, Bohl DG. Separated flow transition mechanism and prediction with high and low freestream turbulence under low pressure turbine conditions. In: Proceedings of ASME Turbo ExpoPower for Land, Sea, and Air, Vienna, Austria. Volume 4: Turbo Expo 2004. ASMEDC; 2004. pp. 45-55

[35] Roberts SK, Yaras MI. Large-Eddy simulation of transition in a separation bubble. Journal of Fluids Engineering. 2005;**128**(2):232-238. DOI: 10.1115/1.2170123

[36] McAuliffe BR, Yaras MI. Numerical study of instability mechanisms leading to transition in separation bubbles. Journal of Turbomachinery. 2008;**130**(2):1-8. DOI: 10.1115/1.2750680

[37] Marxen O, Lang M, Rist U. Discrete linear local eigenmodes in a separating laminar boundary layer. Journal of Fluid Mechanics. 2012;**711**:1-26. Available from: https://www.cambridge.org/core/article/discrete-linear-local-eigenmodes-in-a-separating-laminar-boundary-layer/C44F7EAF1884566E188A26D5D25DFA69

[38] Maucher U, Rist U, Wagner S. Transitional structures in a laminar separation bubble. In: Nitsche W, Heinemann H-J, Hilbig R, editors. New Results in Numerical and Experimental Fluid Mechanics II: Contributions to the 11th AG STAB/DGLR Symposium Berlin, Germany 1998. Wiesbaden: Vieweg+Teubner Verlag; 1999. pp. 307-314. DOI: 10.1007/978-3-663-10901-3_40

[39] Maucher U, Rist U, Wagner S. Secondary disturbance amplification and transition in laminar separation bubbles. In: Fasel HF, Saric WS, editors. Laminar-Turbulent Transition. Berlin, Heidelberg: Springer Berlin Heidelberg; 2000. pp. 657-662

[40] Marxen O, Lang M, Rist U, Wagner S. A combined experimental/numerical study of unsteady phenomena in a laminar separation bubble. Flow, Turbulence and Combustion. 2003;**71**(1-4):133-146

[41] Jones LE, Sandberg RD, Sandham ND. Direct numerical simulations of forced and unforced separation bubbles on an airfoil at incidence. Journal of Fluid Mechanics. 2008;**602**:175-207. Available from: https://www.cambridge.org/core/article/direct-numerical-simulations-of-forced-and-unforced-separation-bubbles-on-an-airfoil-at-incidence/A739F9AEE59B2F240FE33415740342E8

[42] Simoni D, Ubaldi M, Zunino P. Experimental investigation of flow instabilities in a laminar separation bubble. Journal of Thermal Science. 2014;**23**(3):203-214. DOI: 10.1007/s11630-014-0697-3

[43] Lardeau S, Leschziner M, Zaki T. Large Eddy simulation of transitional separated flow over a flat plate and a compressor blade. Flow, Turbulence and Combustion. 2012;**88**(1):19-44. DOI: 10.1007/s10494-011-9353-0

[44] Olson DA, Katz AW, Naguib AM, Koochesfahani MM, Rizzetta DP, Visbal MR. On the challenges in experimental characterization of flow separation over airfoils at low Reynolds number. Experiments in Fluids. 2013;**54**(2):1-11

[45] Lengani D, Simoni D. Recognition of coherent structures in the boundary layer of a low-pressure-turbine blade for different free-stream turbulence intensity levels. International Journal of Heat and Fluid Flow. 2015;**54**:1-13. Available from: https://www.sciencedirect.com/science/article/pii/S0142727X15000326

[46] Simoni D, Lengani D, Ubaldi M, Zunino P, Dellacasagrande M. Inspection of the dynamic properties of laminar separation bubbles: Free-stream turbulence intensity effects for different Reynolds numbers. Experiments in Fluids. 2017;**58**(6):1-14

[47] Istvan MS, Yarusevych S. Effects of free-stream turbulence intensity on transition in a laminar separation bubble formed over an airfoil. Experiments in Fluids. 2018;**59**(3):1-21

[48] Haggmark C. Investigations of Disturbances Developing in a Laminar Separation Bubble Flow (Instability Waves). Stockholm: Royal Institute of Technology; 1987

[49] Li H, Yang Z. Numerical study of separated boundary layer transition under pressure gradient. In: Proceedings of the12th International Conference on Heat Transfer, Fluid Mechanics and Thermodynamics. Spain: Proceedings of the12th International Conference on Heat Transfer, Fluid Mechanics and Thermodynamics; 2016. pp. 1759-1964

[50] Zaki TA, Jang W, Rodi W, Durbin PA. Direct numerical simulations of transition in a compressor cascade: The influence of free-stream turbulence. Journal of Fluid Mechanics. 2010;**665**:57-98. Available from: https://www.cambridge.org/core/article/direct-numerical-simulations-of-transition-in-a-compressor-cascade-the-influence-of-freestream-turbulence/EEFDFE7631794A69966467CC4FC5DA81

[51] Dong Y, Cumpsty NA. Compressor blade boundary layers: Part 2—Measurements with incident wakes. Journal of Turbomachinery. 1990;**112**(2):231-240. DOI: 10.1115/1.2927637

[52] Wissink JG. DNS of separating, low Reynolds number flow in a turbine cascade with incoming wakes. International Journal of Heat and Fluid Flow. 2003;**24**(4):626-635. Available from: https://www.sciencedirect.com/science/article/pii/S0142727X03000560

[53] Palikaras A, Yakinthos K, Goulas A. Transition on a flat plate with a semi-circular leading edge under uniform and positive shear free-stream flow. International Journal of Heat and Fluid Flow. 2002;**23**(4):455-470. Available from: https://www.sciencedirect.com/science/article/pii/S0142727X02001467

[54] Lamballais E, Silvestrini J, Laizet S. Direct numerical simulation of flow separation behind a rounded leading edge: Study of curvature effects. International Journal of Heat and Fluid Flow. 2010;**31**(3):295-306. Available from: https://www.sciencedirect.com/science/article/pii/S0142727X09001726

[55] Yang Z. A comparative study of separated boundary layer transition on a flat plate with a blunt/semi-circular leading edge. In: Proceedings of the 2nd International Conference on Fluid Mechanics, Heat & MassTransfer, CorfuIsland, Greece, July,2011. CorfuIsland, Greece: Proceedings of the 2nd International Conferenceon Fluid Mechanics, Heat & Mass Transfer; 2011.

pp. 191-195. Available from: http://hdl.handle.net/10545/620671

[56] Yang Z. Numerical study of transition process in a separated boundary layer on a flat plate with two different leading edges. WSEAS Trancations on Applied and Theoretical Mechanics. 2012;7(1):49-58

[57] Malkiel E, Mayle RE. Transition in a separation bubble. Journal of Turbomachinery. 1996;118(4):752-759. DOI: 10.1115/1.2840931

[58] Yang ZY, Voke PR. On early stage instability of separated boundary layer transition. In: Advances in Turbulence VIII. Spain: CIMNE; 2000. pp. 145-148

[59] Yang Z, Voke PR. Large-eddy simulation of boundary-layer separation and transition at a change of surface curvature. Journal of Fluid Mechanics. 2001;439:305-333. Available from: https://www.cambridge.org/core/article/largeeddy-simulation-of-boundarylayer-separation-and-transition-at-a-change-of-surface-curvature/3DF401E87D419778AB65FA52264A2992

[60] Yang Z. Large-scale structures at various stages of separated boundary layer transition. International Journal for Numerical Methods in Fluids. 2002;40(6):723-733. DOI: 10.1002/fld.373

[61] Abdalla IE, Yang Z. Numerical study of the instability mechanism in transitional separating–reattaching flow. International Journal of Heat and Fluid Flow. 2004;25(4):593-605. Available from: https://www.sciencedirect.com/science/article/pii/S0142727X04000062

[62] Abdalla IE, Yang Z. Numerical study of a separated-reattached flow on a blunt plate. AIAA Journal. 2005;43(12):2465-2474. DOI: 10.2514/1.1317

[63] Abdalla IE, Cook MJ, Yang Z. Numerical study of transitional separated–reattached flow over surface-mounted obstacles using large-eddy simulation. International Journal for Numerical Methods in Fluids. 2007;54(2):175-206. DOI: 10.1002/fld.1396

[64] Abdalla IE, Yang Z, Cook M. Computational analysis and flow structure of a transitional separated-reattached flow over a surface mounted obstacle and a forward-facing step. International Journal of Computational Fluid Dynamics. 2009;23(1):25-57. DOI: 10.1080/10618560802566246

[65] Gu H, Yang J, Liu M. Study on the instability in separating–reattaching flow over a surface-mounted rib. International Journal of Computational Fluid Dynamics. 2017;31(2):109-121. DOI: 10.1080/10618562.2017.1307968

[66] Vlahostergios Z, Yakinthos K, Goulas A. Experience gained using second-moment closure Modeling for transitional flows due to boundary layer separation. Flow, Turbulence and Combustion. 2007;79(4):361-387. DOI: 10.1007/s10494-007-9103-5

[67] Yang Z, Abdalla IE. Effects of free-stream turbulence on large-scale coherent structures of separated boundary layer transition. International Journal for Numerical Methods in Fluids. 2005;49(3):331-348. DOI: 10.1002/fld.1014

[68] Yang Z, Abdalla IE. Effects of free-stream turbulence on a transitional separated–reattached flow over a flat plate with a sharp leading edge. International Journal of Heat and Fluid Flow. 2009;30(5):1026-1035. Available from: https://www.sciencedirect.com/science/article/pii/S0142727X09000824

[69] Langari M, Yang Z. Numerical study of the primary instability in a separated boundary layer transition under elevated free-stream turbulence. Physics of Fluids. 2013;25(7):074106. DOI: 10.1063/1.4816291

[70] Kotsonis M, Ghaemi S. Forcing
mechanisms of dielectric barrier
discharge plasma actuators at carrier
frequency of 625 Hz. Journal of Applied
Physics. 2011;**110**(11):113301.
DOI: 10.1063/1.3664695

[71] Benard N, Moreau E. Electrical and
mechanical characteristics of surface AC
dielectric barrier discharge plasma
actuators applied to airflow control.
Experiments in Fluids. 2014;**55**(11):
1846. DOI: 10.1007/s00348-014-1846-x

[72] Boutilier MSH, Yarusevych S.
Separated shear layer transition over an
airfoil at a low Reynolds number.
Physics of Fluids. 2012;**24**(8):084105.
DOI: 10.1063/1.4744989

[73] Boutilier MSH, Yarusevych S.
Sensitivity of linear stability analysis of
measured separated shear layers.
European Journal of Mechanics - B/
Fluids. 2013;**37**:129-142. Available from:
https://www.sciencedirect.com/science/
article/pii/S0997754612001057

[74] Marxen O, Dans H. The effect of
small-amplitude convective
disturbances on the size and bursting
of a laminar separation bubble. Journal
of Fluid Mechanics. 2011;**671**:1-33.
Available from: https://www.
cambridge.org/core/article/effect-of-
smallamplitude-convective-
disturbances-on-the-size-and-
bursting-of-a-laminar-separation-bubb
le/5F35CB04F0B6ED7FEDDFC1
04AFECF5CD

[75] Pröbsting S, Yarusevych S. Laminar
separation bubble development on an
airfoil emitting tonal noise. Journal of
Fluid Mechanics. 2015;**780**:167-191.
Available from: https://www.cambridge.
org/core/article/laminar-separation-
bubble-development-on-an-airfoil-
emitting-tonal-noise/76574ABB1894AF
FDADC429B498CB0D1A

The Phenomenon of Friction Resistance Due to Streamwise Heterogeneous Roughness with Modified Wall-Function RANSE

I. Ketut Aria Pria Utama, I. Ketut Suastika
and Muhammad Luqman Hakim

Abstract

Surface roughness can reduce the performance of a system of fluid mechanics due to an increase in frictional resistance. The ship hull, which is overgrown by biofouling, experiences a drag penalty which causes energy wastage and increased emission levels. The phenomenon of fluid flow that passes over a rough surface still has many questions, one of which is the phenomenon of frictional resistance on heterogeneous roughness in the streamwise direction. In the ship hull, biofouling generally grows heterogeneous along the hull with many factors. RANSE-based Computational Fluid Dynamics was used to investigate the friction resistance for heterogeneous roughness phenomenon. The modified wall-function method represented equivalent sand grain roughness (k_s) and a roughness function were applied together with k-epsilon turbulence model to simulate rough wall turbulent boundary layer flow. As the heterogeneous roughness, three different k_s values were denoted as P (k_s = 81.25 μm), Q (k_s = 325.00 μm) and R (k_s = 568.75 μm), and they are arranged by all possible combinations. The combined roughness, whether homogeneous (PPP, QQQ, or RRR) and inhomogeneous (PQR, PRQ, QPR, etc.), results in unique skin friction values. The step-change in the height of the hetero-geneous roughness produced a sudden change in the local skin friction coefficient in the form of overshoot or undershoot, followed by a relaxation where the inhomo-geneous local skin friction is slowly returning to the homogeneous local one, which was explained in more detail by plotting the distribution of the mean velocity profile near the step-up or step-down. The order of roughness arrangement in a streamwise heterogenous roughness pattern plays a key role in generating overall skin friction with values increasing in the following order: PQR < PRQ < QPR < QRP < RPQ < RQP. Those inhomogeneous cases with three different values of k_s can be represented by a single value (being like homogeneous) by the calculations provided in this paper.

Keywords: Heterogeneous roughness, Inhomogeneous roughness, RANSE simulations, Skin friction, modified wall-function

1. Introduction

The issue of using energy more efficiently on ships seems urgent, and how to do this is greatly helped by the existence of CFD. In the last three decades, CFD as a numerical method, which is very sophisticated that can help humans to solve various problems in the fields of science and technology [1–6], but at a competitive cost, has played a very important role in advancing transportation technology [7], especially in naval architecture [8]. With this highly sophisticated method, the need for increased energy efficiency on ships is greatly helped. In 2012, the International Maritime Organization (IMO) noted that the total emissions from the shipping sector worldwide were 2.2% compared to all human-made CO_2 emissions [9]. This number was predicted to increase 2–3 times in 2050 if there are no prevention efforts [10]. There are many ways to save on the use of energy on board [11, 12], such as improving hull and propeller design for more hydrodynamic performance. With the help of CFD, efforts to improve energy efficiency can be made more accessible.

Caring about the cleanliness of the hull due to biofouling is one of the efforts to maintain the hydrodynamic performance to prevent energy waste. Roughness can increase friction resistance, and then power requirements increase, resulting in losses, which have a significant impact on large vessels such as the VLCC (Very Large Crude Carrier) [13] or ships with low Fr (Froude number) [14]. A roughness, namely tubeworm fouling, can increase the friction resistance of ships by 23–34% [15], while a heavily fouled ship hull can increase the friction resistance by up to 80% [16]. Due to the growth of biofouling on ship hull, fuel consumption can increase over the operational time and can increase significantly just in a year [17]. The total economic losses from biofouling, including fuel additions, cleaning, and repainting, can reach $ 15 million a year [18].

The phenomenon of the effect of roughness on fluid flow was first investigated by Nikuradse [19]. The mean velocity profile of the structure turbulence boundary layer of the smooth case (see Eq. (1)) is exposed to a downward shift in the log law region by a roughness to become a new velocity profile (see Eq. (2)) [20]. Thus, the concept of the roughness function (ΔU^+) as the downward shift and the roughness Reynold number (k_s^+) were used (see Eqs. 3 and 4). Where: U^+ is the non-dimensional mean velocity profile equal to U/U_τ; U is the mean velocity at y (the normal from the wall); U_τ is the friction velocity defined as $\sqrt{\tau_w/\rho}$; τ_w is the shear stress magnitude and ρ is the density of the fluid; y^+ is the non-dimensional normal distance from the wall defined as yU_τ/ν; ν is the kinematic viscosity; κ is the von Karman constant and B is the smooth wall log-law intercept; k_s is equivalent sand roughness height. From the new velocity profile, there is an indication of an increased momentum deficit compared to the smooth case.

$$U^+{}_{smooth} = \frac{1}{\kappa} \ln y^+ + B \tag{1}$$

$$U^+{}_{rough} = \frac{1}{\kappa} \ln y^+ + B - \Delta U^+ \tag{2}$$

$$\Delta U^+ = f\left(k_s^+\right) \tag{3}$$

$$k_s^+ = \frac{k_s U_\tau}{\nu} \tag{4}$$

The flow over surface roughness phenomenon can be simulated using CFD, which is generally solved by two different methods: modified wall function and geometrically resolved. The modified wall function is a method in which the geometry model (mesh) remains smooth, and the roughness length scale represents the roughness, generally using k_s (equivalent sand grain roughness height) as a variable for a roughness function (ΔU^+) which will shift/modify the mean velocity profile. This method is only supported when using the RANSE (Reynolds-Averaged Navier–Stokes Equations). The modified wall function method is very effective for modeling large objects such as the hull of a ship, where the k_s and ΔU^+ values have been previously known and inputted, as was done by Demirel et al. [21], Song et al. [22], Andersson et al. [23], and also it can be used for propeller [24] and tidal turbine [25]. The modified wall function method prioritizes seeing results (impacts), such as increased drag, wake, and the like. Meanwhile, the geometrically resolved method used the real roughness geometry that formed from the mesh. These methods generally use to know how the k_s value and the ΔU^+ characteristic from a roughness. DNS (Direct Numerical Simulation) and LES (Large Eddy Simulation) are well known to be very good at doing this task. The geometrically resolved can also be done with RANSE as done by Atencio & Chernoray [26] with a difference of about 7% with their experimental results.

A reduction in hull performance due to roughness can be simulated using CFD with a modified wall function method with acceptable accuracy. Reynolds-averaged Navier–Stokes Equations (RANSE) simulation to study the friction resistance of flat plates due to the antifouling coating performed by Demirel et al. [27]. They used the roughness function from the experimental result of Schultz [28]. Using the Kriso Container Ship model, Demirel et al. [21] continued the CFD simulation to predict the impact of marine coatings and biofouling. Song et al. [22] also looked at the effect of biofouling on the ship's hydrodynamic characteristics, using a different roughness function. Anderson et al. [23] performed a comprehensive review and comparative analysis of different methods to model hull roughness. The comparison of acceptable CFD and experimental results was carried out by Song et al. [29].

Much of the literature reported assumes that the roughness distribution is homogeneous, but in reality, it is much non-homogeneous. Hull roughness, mainly that arising from biofouling, rarely occurs homogeneously. From a personal review in the field, the authors found that the biofouling growth was thicker in the stern of the vessel than in the bow, which may be influenced by the distribution of shear stress and flow compressive forces, which are more favorable for biofouling to grow better at the stern [30].

Many studies are looking at the roughness problems at an inhomogeneous pattern. The smooth-to-rough and rough-to-smooth patterns in streamwise direction were studied by Antonia & Luxton [31, 32]. In a streamwise phenomenon, the flow of fluid through abruptly different roughness conditions produces an internal boundary layer, which limits the near-wall layer, which senses the new surface conditions, from the flow further away from the wall, which keeps a memory of the upstream surface conditions before the surface transition [33]. The internal boundary layer and the local wall shear stress, at the transition of the two difference roughness conditions, exceed the equilibrium value when the roughness is homogeneous and then change to relax according to the homogeneous roughness equilibrium value certain distance [34]. Experimental methods and numerical simulations carried out several studies related to this inhomogeneous roughness. However, the numerical simulations for solving this case are mostly DNS and LES. Few do that through the RANSE, as Suastika et al. [35] on a flat plate, and Song et al. [36] using the Wigley hull model.

The frictional resistance acting on the hull due to inhomogeneous roughness becomes important to be modeled and analyzed, considering that the hull roughness due to biofouling is mostly inhomogeneous. In a simple method where the ship hull is represented by a flat plate, then an inhomogeneous roughness is applied by dividing the plate into three equal parts, namely the fore, middle, and after. The parts are given different roughness values in the form of k_s and are arranged according to several combinations. From the simulation results on inhomogeneous roughness in the form of frictional resistance values will be compared with where if the condition is a smooth surface and some surfaces with homogeneous roughness. Then how are the three roughness values in the inhomogeneous condition correlated to become one roughness value (homogenized) which is close to the inhomogeneous roughness value. This CFD simulation uses the basis of Reynolds-averaged Navier–Stokes Equations (RANSE), where the roughness model uses a modified wall function. Research with variations in roughness that is streamwise inhomogeneous, which is then analyzed systematically, according to our knowledge is still a little done.

2. CFD modeling

A Reynolds-averaged Navier–Stokes Equations (RANSE) model, implemented in ANSYS Fluent, was used to solve the governing equations that can model turbulent flow over rough walls using modified wall function. The governing equations consist of averaged continuity and momentum equations, which for an incompressible flow without body forces, are given using tensor notation as described in Eqs. 5 and 6, respectively. Where $i, j = 1, 2, 3$, U_i is the mean velocity component, P is the mean pressure, ρ is the fluid density, ν is the kinematic viscosity, u_i' is the fluctuating velocity component, and $\overline{u_i' u_j'}$ is the Reynolds stresses [37].

$$\frac{\partial U_i}{\partial x_i} = 0 \tag{5}$$

$$\frac{\partial U_i}{\partial t} + \frac{\partial (U_i U_j)}{\partial x_j} = -\frac{1}{\rho}\frac{\partial P}{\partial x_i} + \frac{\partial}{\partial x_j}\left[\nu\left(\frac{\partial U_i}{\partial x_j} + \frac{\partial U_j}{\partial x_i}\right)\right] - \frac{\partial\left(\overline{u_i' u_j'}\right)}{\partial x_j} \tag{6}$$

The realizable k–ε with standard wall function turbulence model, which relates the Reynolds stresses to the mean flow properties, was used to close the system of Eqs. (5) and (6). The turbulence model is a two-equation model representing the transports of turbulence kinetic energy k and turbulence dissipation rate ε [38].

The roughness function (ΔU^+) model used in this study is that proposed by Cebeci and Bradshaw [39], whose model follows Nikuradse's uniform sand-grain roughness data [19]. Therefore, the roughness height utilized in this study is referred to as equivalent sand-grain roughness height k_s. The generalized Cebeci and Bradshaw's roughness function model is given in Eq. (7), where $A = 0$, $k_{s;smooth}^+ = 2.25$, $k_{s;rough}^+ = 90.00$ and $C_s = 0.253$. In which the power a is given in Eq. (8).

$$\Delta U^+ = \begin{cases} 0 & \rightarrow k_s^+ \leq k_{s;smooth}^+ \\ \frac{1}{\kappa}\ln\left[A\left(\frac{k_s^+/k_{s;smooth}^+}{k_{s;rough}^+ - k_{s;smooth}^+}\right) + C_s k_s^+\right]^a & \rightarrow k_{s;smooth}^+ < k_s^+ \leq k_{s;rough}^+ \\ \frac{1}{\kappa}\ln\left(A + C_s k_s^+\right) & \rightarrow k_s^+ > k_{s;rough}^+ \end{cases} \tag{7}$$

$$a = \sin \left[\frac{\pi}{2} \frac{\log \left(k_s^+ / k_{s;smooth}^+ \right)}{\log \left(k_{s;rough}^+ / k_{s;smooth}^+ \right)} \right] \qquad (8)$$

The boundary conditions were set for this study, as illustrated in **Figure 1a**. The inlet was velocity inlet where the free stream velocity prescribes the flow velocity U_∞. The outlet was the pressure outlet set to be hydrostatic to ensure no upstream propagation of disturbances. The no-slip condition is applied on the plate's surface while the top and side boundaries are modeled as free-slip walls. The boundary conditions, governing, and turbulence modeling equations are discretized using a finite volume second-order method. These sets are then solved using a finite volume solver, utilizing a SIMPLE algorithm in which gradient calculations are carried out using the least-square cell-based method. The residual is set at 10^{-5} as a convergence criterion. For all simulations in this study, the plate length, fluid properties, and free stream velocity are kept constant. The plate length L is 30 m. The fluid is seawater with mass density ρ = 1025 kg/m^3 and dynamic viscosity μ = 0.001077 kg/(ms). Finally, the free stream velocity is set at U_∞ = 9.77 m/s (19 knots).

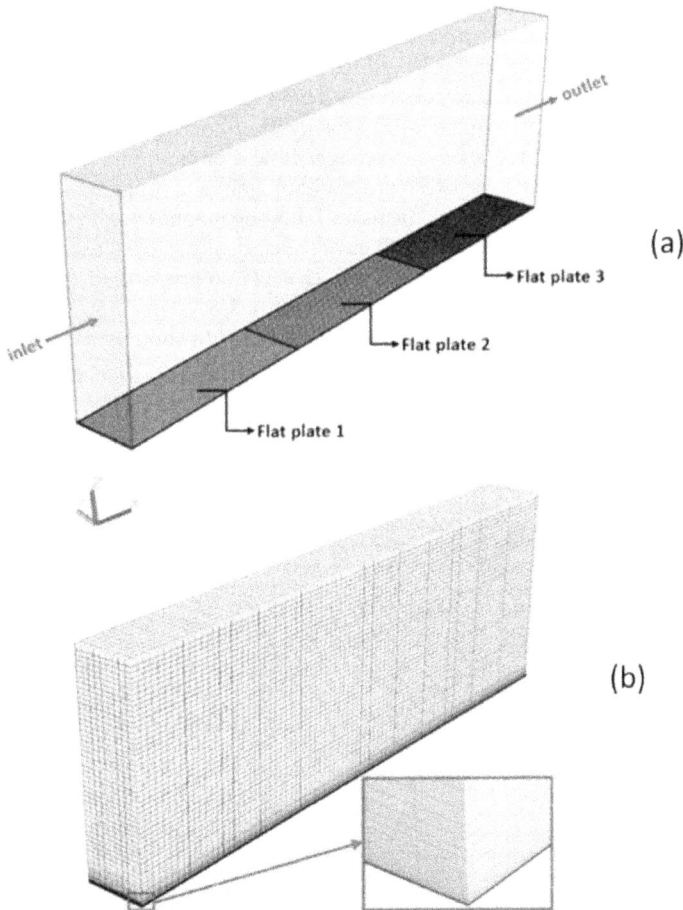

Figure 1.
The flow domain used in the simulations with three plate segments at the bottom boundary (a), and the hexahedron mesh with an exponential cell height gradation near the wall bottom boundary (b).

In CFD modeling, the distance of the frame against the wall is set to decrease exponentially as it moves typically towards the wall., as shown in **Figure 1b**. A hexahedron-type mesh is chosen, and it is arranged manually with adjustable grid size. It is crucial to determine the mesh size near the wall to obtain an appropriate value for the dimensionless normal coordinate y^+, defined as $y^+ = u\tau y/v$, where y is the outward wall-normal coordinate. To model the roughness effects correctly, the y^+ value for the first cell center above the wall must be larger than the local roughness Reynolds number k_s^+. To ensure this condition is always satisfied, ANSYS Fluent will virtually shift the wall if $y^+ < k_s^+$. For the roughness cases considered in this study, a blockage effect of 50% of the roughness height is assumed, and the corrected y^+ value for the first cell center above the wall is given as $y^+ = y^+ + k_s^+/2$. In this way, the singularity issue is avoided, and fine meshes can be handled correctly.

3. Surface roughness modeling

In this study, we will investigate just single parameter variations, namely the roughness height k_s. This section will explain details of the k_s set up and their possible combinations. Four surface roughness with different k_s values are considered in this study, namely, smooth wall (S), small roughness height (P), medium roughness height (Q), and high roughness height (R). All three combinations of roughness P, Q, and R are considered to form either homogeneous or inhomogeneous rough walled turbulent boundary layer flow. For example, a three-surface combination of PPP, QQQ, or RRR forms a homogeneous roughness, while a combination of PQR, PRQ, QPR, etc., forms an inhomogeneous roughness, where those are described in **Figure 2**. These k_s values are specifically chosen so that the average height of three different k_s of P = 81.25 μm, Q = 325.00 μm, and R = 568.75 μm will give an average height equal of Q, i.e., (81.25 μm + 325.00 μm + 568.75 μm)/ 3 = 325.00 μm. The selected k_s values are also designed to simulate the various stages of ship-hull biofouling growth, ranging from light slime [16] to about small calcareous fouling [15].

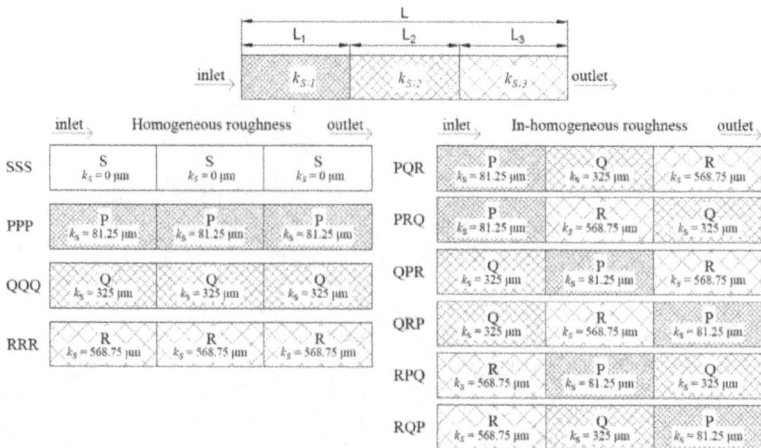

Figure 2.
Combinations of three plate segments resulting inhomogeneous (including fully smooth) and inhomogeneous rough surface conditions.

4. Verification and validation of modeling results

4.1 Grid independence study

To ensure optimum numbers of cells are used in the final simulations, several grid independence tests using flat plate smooth wall (SSS) base cases are conducted. For each case, the overall friction coefficient C_F is calculated using the increasing number of cells in the simulation. The number of cells in the latter simulation is approximately twice than that in the former. The overall friction coefficient C_F is defined in Eq. (9), where D is the drag per unit width, τ_w is the wall shear stress, ρ is the fluid density, U_∞ is the free stream velocity, L is the plate length, and x is the distance downstream from the leading edge of the plate. Furthermore, a percent error $e_{n+1,n}$ between the lower and higher cell numbers is defined in Eq. (10) [40].

$$C_F = \frac{D}{\rho U_\infty^2 L/2} = \frac{\int_0^L \tau_w dx}{\rho U_\infty^2 L/2} \tag{9}$$

$$e_{n+1,n} = \frac{C_F(n+1) - C_F(n)}{C_F(n)} \times 100\% \tag{10}$$

The results of the grid independence study are summarized in **Table 1**. The number of cells is varied from 750,000 to 6,127,550. **Table 1** shows that the C_F value increases monotonically with the increasing number of cells, which is expected to reach an asymptotic value at a very large number of cells. The value of $e_{n+1,n}$ as listed in **Table 1** decreases with the increasing number of cells used in the simulation. The results show that the error is very low (in the range 0.0206–0.131%), well below the recommended 2% from the literature [41]. Based on this grid independence test, N = 3,000,000 is chosen as an optimum number of cells for all the cases (including homogeneous and inhomogeneous roughness).

4.2 Verification and validation

In addition to the grid independence tests, further analyses are carried out for varied viscous-scaled wall-normal distance y^+ ranges of the first cell center above the wall. For the smooth plate SSS case, the calculation result is verified using the well-known Schoenherr's friction coefficient and the 1957 ITTC (International Towing Tank Conference) ship-model correlation line. Schoenherr's friction coefficient C_F is given in Eq. (11). It was adopted by ATTC (American Towing Tank Conference) as a standard for the clean hull skin friction resistance in 1947, and it is often referred to as the 1947 ATTC line. The second correlation is the 1957 ITTC ship-model correlation line, which is given in Eq. (12). A percent error is defined

Run number n	Number of cells N	$C_F \times 10^3$	Percent error $e_{n+1,n}$ [%]
1	500,000	1.788	—
2	1,522,950	1.790	0.1310
3	3,000,000	1.791	0.0332
4	6,127,550	1.791	0.0206

Table 1.
Friction coefficients C_F calculated using increasing number of cells in the simulations for the smooth plate (SSS).

Case	Re_L	$C_{F;1}$	$C_{F;2}$	$C_{F;3}$	C_F	$e_{r;s}$	$e_{i;q}$	$e_{i,h}$
		$\times 10^3$	$\times 10^3$	$\times 10^3$	$\times 10^3$	[%]	[%]	[%]
H30_SSS	2.79×10^8	2.075	1.700	1.596	1.790	0.00		
H30_PPP	2.79×10^8	2.608	2.062	1.915	2.195	22.60		
H30_QQQ	2.79×10^8	3.436	2.660	2.455	2.850	59.19	0.00	
H30_RRR	2.79×10^8	3.809	2.918	2.685	3.137	75.23		
I30_PQR	2.79×10^8	2.607	2.736	2.759	2.700	50.82	−5.26	1.87
I30_PRQ	2.79×10^8	2.607	3.037	2.473	2.706	51.12	−5.07	1.71
I30_QPR	2.79×10^8	3.437	2.007	2.770	2.738	52.92	−3.94	0.37
I30_QRP	2.79×10^8	3.436	2.951	1.846	2.745	53.29	−3.71	−0.44
I30_RPQ	2.79×10^8	3.811	1.988	2.484	2.761	54.21	−3.13	−0.52
I30_RQP	2.79×10^8	3.810	2.633	1.846	2.763	54.32	−3.06	−1.14

Table 2.
Overall friction coefficients for the plate segments $C_{F;1}$, $C_{F;2}$, $C_{F;3}$ and the entire plate C_F, and the percent errors $e_{r;s}$, $e_{i;q}$ and $e_{i,h}$.

Case	Re_L	y^+ range		$C_F \times 10^3$			Percent error [%]	
		Min	Max	CFD	ATTC'47	ITTC'57	ATTC'47	ITTC'57
SSS	2.79×10^8	64	112	1.773	1.802	1.805	−1.56	−1.76
SSS	2.79×10^8	155	254	1.791	1.802	1.805	−0.62	−0.62

Table 3.
Overall friction coefficient C_F calculated using different ranges of y^+ value for the first cell center above the wall compared with 1947 ATTC (Eq. (11)) and 1957 ITTC (Eq. (12)) lines for the smooth SSS case.

between the CFD result and the 1947 ATTC line and, in a similar manner, between the CFD result and the 1957 ITTC ship-model correlation line to quantify the accuracy of the CFD results. The percent error e in the latter case is calculated using Eq. (13).

$$\frac{0.242}{\sqrt{C_F}} = \log_{10}(Re\, C_F) \tag{11}$$

$$C_F = \frac{0.075}{\left[\log_{10}(Re) - 2\right]^2} \tag{12}$$

$$e = \frac{C_{F;CFD} - C_{F;ITTC\ 1957}}{C_{F;ITTC\ 1957}} \times 100\% \tag{13}$$

The results are summarized in **Table 3**, showing C_F values calculated using different y^+ ranges for the first cell center above the wall targeted in the simulations. **Table 3** shows that for the SSS case, using a y^+ range between 155 and 254 would result in the closest C_F value to the 1947 ATTC and 1957 ITTC lines with percent errors of $e = 0.620\%$ and $e = 0.814\%$, respectively. A smaller y^+ range will result in larger differences, but the percentage differences do not exceed 1.76%, which is relatively small. Despite these anomalies, overall, the CFD and the 1947 ATTC or the 1957 ATTC line are small. Despite some discrepancies, for the smooth surface cases considered in this study, any y^+ range between 64 and 254 will result in an acceptable C_F value with a maximum magnitude of percent errors of 1.57%

when compared with the 1947 ATTC line and 1.76% when compared with the 1957 ITTC line. This result is following the recommended y^+ range in the literature for smooth flat plate CFD simulations between 50 and 300 [42].

To model the roughness effects correctly, the y^+ value for the first cell center above the wall denoted as $(\Delta y^+)_1$, must be larger than the local equivalent sand grain roughness Reynolds number k_s^+, i.e., $(\Delta y^+)_1 > k_s^+$. However, when one employs a fine mesh near the wall, the $(\Delta y^+)_1$ value may have a smaller value than the k_s^+ value, i.e., $(\Delta y^+)_1 < k_s^+$. If such a case happens, ANSYS Fluent applies a virtual shift of the wall by increasing the value of $(\Delta y^+)_1$ with an amount $k_s^+/2$, such that $(\Delta y^+)_1 > k_s^+$.

To gain more insight into the CFD results for rough conditions, they are verified using the empirical calculation, Granville's similarity law scaling method [43]. The simplified Granville similarity scaling can be calculated using Eqs. 14, 15, and 16. Where: C_{F_R} is the coefficient of frictional resistance for rough condition, where the empirical formula as the foundation is taken from the approximated Kármán-Schoenherr formula [44]; Re_r is the Reynolds number for calculate the C_{F_R} using the empirical formula, which equal of the Reynolds number for smooth condition (Re_s) that is shifted as described in Eq. (15). Then, κ is the von Kármán constant; k_s^+ is roughness Reynolds number; ν is kinematic viscosity; U_τ is friction velocity defined as $\sqrt{\tau_w/\rho}$ or approached by $U_\infty(C_F/2)^{1/2}$; τ_w is the shear stress magnitude, where it is necessary to do iterative calculations against C_{F_R}. The roughness function ΔU^+ is from Cebeci and Bradshaw [39] in Eq. (7), according to this study.

The verification result using the similarity scaling from Granville [43] can be seen in **Table 4**. The calculation uses $e_{C;G}$, as described in Eq. (17). From the results of the calculation of $e_{C;G}$, it can be concluded that CFD modeling for homogeneous roughness can be accepted with the difference in error against the empirical is not exceed 1.8%.

$$C_{F_R} = \frac{0.0795}{\left(Log_{10} Re_r - 1.729\right)^2} \quad (14)$$

$$Re_r = Re_s - 10^{\left(\frac{\Delta U^+ + \kappa}{\ln(10)}\right)} \quad (15)$$

$$\Delta U^+ = f(k_s^+) = f\left(\frac{k_s U_\tau}{\nu}\right) \quad (16)$$

$$e_{C;G} = \frac{C_{F;CFD} - C_{F;Granville}}{C_{F;Granville}} \times 100\% \quad (17)$$

$$k_s^+ = \left(\frac{k_s}{L}\right)\left(\frac{Re\, C_{FS}}{2}\right)\left(\sqrt{\frac{2}{C_{FR}}}\right)\left[1 - \frac{1}{\kappa}\left(\sqrt{\frac{2}{C_{FR}}}\right) + \frac{1}{\kappa}\left(\frac{3}{2\kappa} - \Delta U^{+'}\right)\left(\frac{C_{FR}}{2}\right)\right] \quad (18)$$

Case	Re_L	$C_F \times 10^3$		$e_{C;G}$ [%]
		CFD	Granville	
PPP	2.79×10^8	2.195	2.191	0.20
QQQ	2.79×10^8	2.850	2.880	−1.03
RRR	2.79×10^8	3.138	3.195	−1.79

Table 4.
Overall friction coefficient C_F of homogeneous rough condition calculated compared with the similarity law scaling procedure from Granville [43].

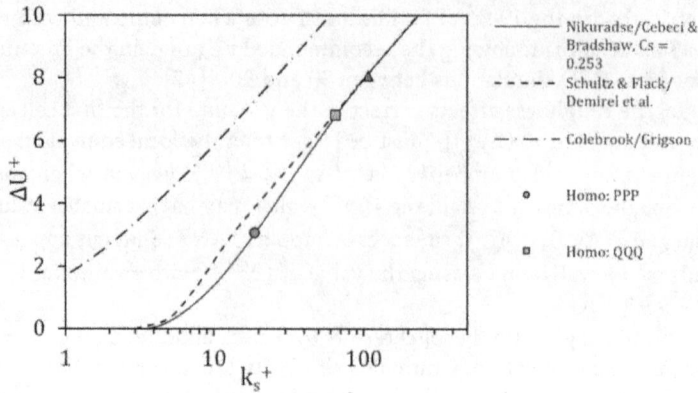

Figure 3.
ΔU^+ and k_s^+ for homogeneous cases that calculated using Granville [45] compared with the used roughness function, and other roughness functions, Colebrook-type roughness function [46], and Schultz and Flack [47].

$$\Delta U^+ = \left(\sqrt{\frac{2}{C_{FS}}}\right) - \left(\sqrt{\frac{2}{C_{FR}}}\right) - 19.7\left[\left(\sqrt{\frac{C_{FS}}{2}}\right) - \left(\sqrt{\frac{C_{FR}}{2}}\right)\right] - \frac{1}{\kappa}\Delta U^{+\prime}\left(\sqrt{\frac{C_{FR}}{2}}\right)$$

(19)

The homogeneous roughness simulation results were also verified using other literature from Granville [45], as written in Eqs. 18 and 19. This method can predict the characteristic roughness function, $\Delta U^+ = f(k_s^+)$, by plotting the predicted value of ΔU^+ (Eq. (19)) against k_s^+ (Eq. (18)) with the difference in the overall drag results from the rough conditions (C_{FR}) to the smooth conditions (C_{FS}). Where, L is the plate length, Re is the Reynolds number, k_s is the roughness height, and $\Delta U^{+\prime}$ is the roughness function slope, which is the slope of ΔU^+ as a function of $\ln(k_s^+)$. The verification results are plotted in **Figure 3** and the simulation results successfully approached the planned roughness function model, namely from Cebeci and Bradshaw [39], with $C_s = 0.253$. Verification of the simulation results using Granville [45] method described in **Figure 3**. The results collapse on the roughness function used (Cebeci and Bradshaw [39]).

5. Results and discussion

A systematic analysis of the results from the inhomogeneous rough surface cases is given in this section. To study the roughness effects, the local (c_f) and overall (C_F) skin friction coefficients are calculated for both the homogeneous and inhomogeneous roughness cases. The effects from the roughness height and the roughness sequence in the streamwise direction are studied by analyzing the plots of local skin friction coefficient as a function of the length and plotting the mean velocity profile for the step up and step-down phenomenon and by comparing its integral values (C_F) for the different cases. We also calculate the skin friction coefficient percentage differences between rough surfaces (both homogeneous and inhomogeneous) and the smooth wall reference case and between inhomogeneous roughness cases (combination of PQR) and the homogeneous roughness reference case (i.e., QQQ). Lastly, we carried out the prediction of how the single k_s value of the homogeneous case that equal to the three different k_s that composed the inhomogeneous case.

5.1 Local skin friction c_f

The local skin friction coefficient c_f is defined in Eq. (20). Where τ_w is the wall shear stress (obtained from CFD simulation), ρ is the fluid density and U_∞ is the free stream velocity. The streamwise length x and lateral position y for the inhomogeneous RPQ case are plotted in **Figure 4**. Generally, c_f is plotted against Re_x, but in this case, Re_x is represented by x (streamwise distance) because we want to study the step-up and step-down phenomena. The factors of streamwise length (L) and freestream velocity (U_∞), which are components of the Re_x value, have different effects on the increase in friction resistance [48].

$$c_f = \frac{\tau_w}{\rho U_\infty^2/2} \qquad (20)$$

5.1.1 Homogeneous and inhomogeneous roughness

Figure 4 shows that the rough-wall homogeneous cases have a higher c_f than that of the smooth wall case at the same position on the streamwise. This value indicates that a rough wall surface indeed deviates from the smooth wall case and increases skin friction drag [19]. Within the homogeneous rough wall cases, the plots show that the highest k_s case (RRR) has a higher c_f value than those of the lower k_s cases (Q and P respectively) at equal place. Such behavior shows that a rougher surface will experience an elevated wall drag compared with less rough surfaces. The four homogeneous cases (including the smooth wall case) show a similar monotonic decrease in c_f with increasing x. This classical result of exponential decrease of c_f with x or Re_x illustrates large friction near the leading edge of the plate (low Reynolds numbers), which decreases exponentially towards the trailing edge. Similar behavior has been reported in various experimental and numerical studies [15, 49].

For the inhomogeneous cases with step changes in the equivalent sand grain roughness height k_s (PQR, PRQ, QPR, QRP, RPQ and RQP), the c_f values show step responses following the step-change in k_s. For example, **Figure 4a** with PQR step changes case show an increase in c_f every time there is an increase from P to Q and from Q to R height. **Figure 4a** shows that in the first one-third part of the inhomogeneous rough plate (0 m $< x <$ 10 m) with k_s value of 81.25 mm (P), the inhomogeneous case line (solid red line) collapses with the homogeneous PPP case (yellow dotted line) well. However, as the inhomogeneous case arrives at the start of the second one-third part of the plate (10 m $< x <$ 20 m), where it has k_s value of 325.00 μm (Q), the red line slightly jumps over (overshoots) the homogeneous QQQ case represented with a green dashed-dotted line, and then the red line gradually falls onto the homogeneous QQQ case. Finally, the last one-third part of the inhomogeneous rough plate (20 m $< x <$ 30 m) has k_s value of 568.75 μm (R), and the plot clearly shows that the red line slightly overshoots the homogeneous RRR case (dashed blue line) in the first few x and then gradually collapses to the homogeneous RRR case. Similar behavior is observed for all of the other five inhomogeneous roughness combinations (**Figure 4b–f**).

Such a jump in c_f, Andreopoulos and Wood [50] have reported values between one surface profile to another surface profile. They measured the response of a smooth wall boundary layer to a perturbation/disturbance caused by a short sandpaper strip. The measured τ_w behind the strip was around three times the undisturbed value (fully smooth case). The sudden jump in c_f is followed by a

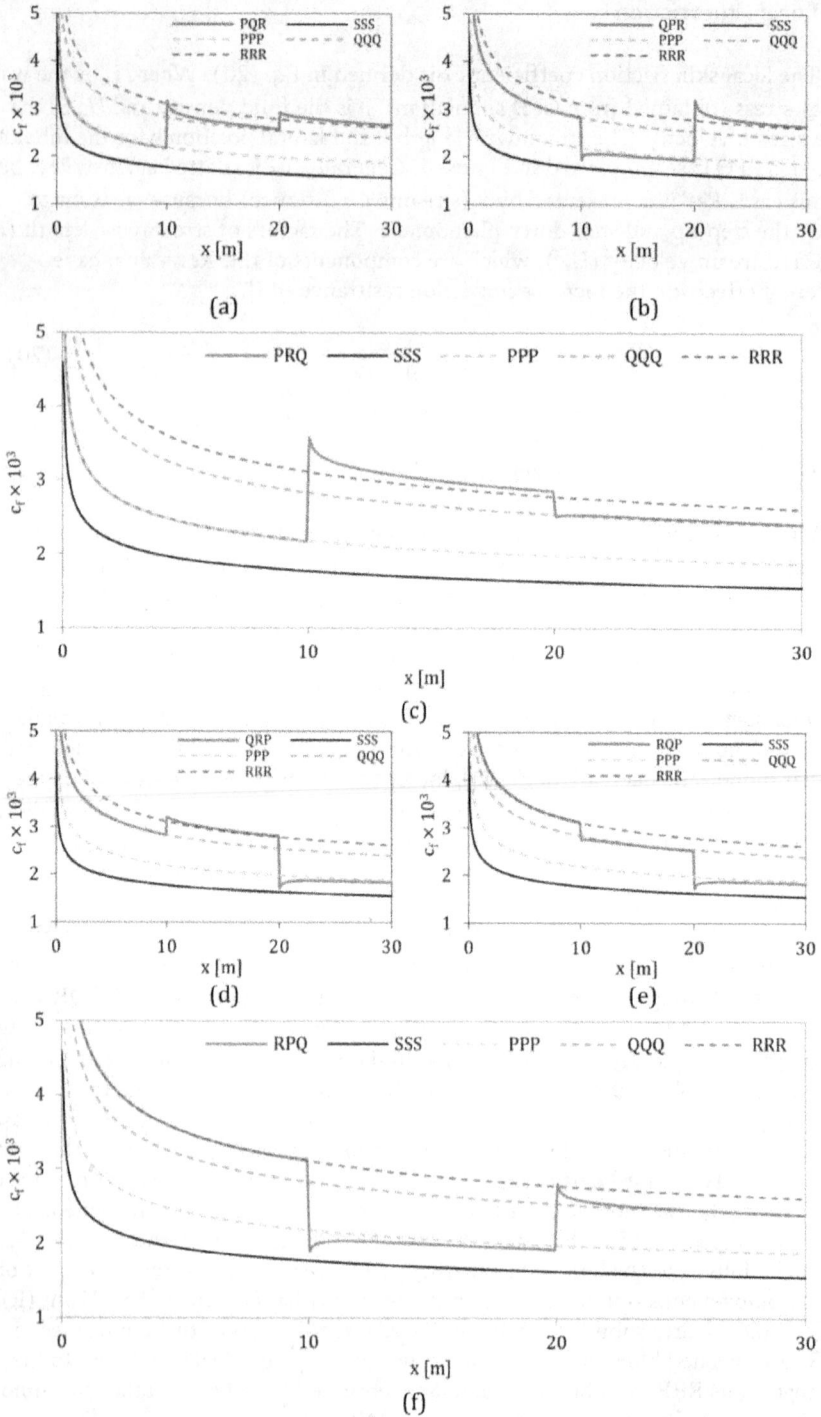

Figure 4.
The distribution of local skin friction coefficient c_f along the plate length for the inhomogeneous PQR (a), PRQ (c), QPR (b), QRP (d), RPQ (f), and RQP (e) cases compared with the homogeneous SSS, PPP, QQQ, and RRR cases.

relaxation, where the c_f is slowly returning to the smooth wall value. The relaxation rate was found to be very slow and Andreopoulos and Wood [50] were unable to record any full recovery, even at the last measuring point. As observed by Andreopoulos and Wood [50], our CFD results of local c_f shown in Figure 9 also exhibit a slow relaxation rate. However, RANSE cannot pick up the small-scale turbulence structures near the wall that occur at the border between the two roughness zones, influencing the flow downstream. The wall model cannot fully capture the flow physics, but it provides us with some indications of the effect. **Figure 4** indicate that the c_f value of the inhomogeneous rough surface will recover the underlying homogeneous rough wall c_f further downstream if the distance is sufficiently long. To quantify this, an averaged overshoot/undershoot will be defined and calculated in the following sub-subsection.

To see what happens to the overshoot and undershoot phenomena, the mean velocity profile of the difference c_f values are plotted in **Figure 5**. The mean velocity profile plot was taken at a distance of 10.25 m from the leading edge. The step-up case where the c_f value overshoot was taken in the PRQ (see **Figure 4c**) and RRR cases, while the step-down case, where the c_f value was undershot, was taken in the RPQ (see **Figure 4f**) and PPP cases. The outer scaling method is used to compare the mean velocity of the profile, where y is the vertical distance from the wall, δ is the thickness of the boundary layer taken 0.99 U_∞, U_∞ is the free stream velocity, and U is the velocity at each y. The plot results show that in the roughness step-up where overshoot occurs, the velocity profile is shifted upward (see **Figure 5a**). Conversely, in the roughness step-down, where there is an undershoot of the cf. value, the velocity profile is shifted downwards (see **Figure 5b**).

5.1.2 Overshoot and undershoot percentage differences

The overshoot and undershoot height of the flow seems to be based on the k_s of the following roughness. For example, when we look into the cases PQR and PRQ in the first row of **Figure 5a** and **5b**, the jump from P to Q is lower than that from P to R, resulting in a lower overshoot from P to Q than that from P to R. This also leads to a faster settling time for the P to Q jump than that for P to R case. Such behavior happens because R corresponds to a much higher k_s value than that corresponds to the Q case. Such undershoot and overshoot raise a question regarding how much is

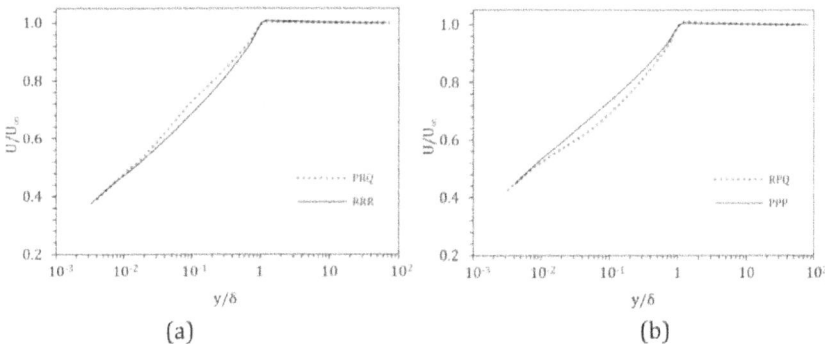

Figure 5.
Comparison of the mean velocity profile plot at x = 10.25 m from the leading edge with the outer scaling method to see the overshoot phenomenon for the step-up roughness (a), and undershoot for the step-down roughness (b).

the difference in c_f between the homogeneous and inhomogeneous cases. To answer such a question, a percent error $e_{i,h}$ is defined between the areas under the c_f curves for the inhomogeneous and homogeneous cases described in Eq. (21), where the subscripts h and i refer to homogeneous and inhomogeneous cases, respectively. The integral boundaries and the surface roughness for the homogeneous and inhomogeneous cases correspond to each other. For example, $e_{i,h}$ for the inhomogeneous case QPR is calculated as described in Eq. (22).

$$e_{i,h} = \frac{\int (c_{f;i} - c_{f;h}) \, d\,\mathrm{Re}_x}{\int (c_{f;h}) \, d\,\mathrm{Re}_x} \times 100\% \qquad (21)$$

$$e_{i,h} = \frac{\int_0^{L_1} (c_{f;QPR} - c_{f;QQQ}) \, d\,\mathrm{Re}_x + \int_{L_1}^{L_1+L_2} (c_{f;QPR} - c_{f;PPP}) \, d\,\mathrm{Re}_x + \int_{L_1+L_2}^{L_2+L_3} (c_{f;QPR} - c_{f;RRR}) \, d\,\mathrm{Re}_x}{\int_0^{L_1} c_{f;QQQ} \, d\,\mathrm{Re}_x + \int_{L_1}^{L_1+L_2} c_{f;PPP} \, d\,\mathrm{Re}_x + \int_{L_1+L_2}^{L_2+L_3} c_{f;RRR} \, d\,\mathrm{Re}_x} \times 100\%$$

$$(22)$$

Eq. (21) shows that if the boundary layer responded to the step-change instantly and there is no overshoot/undershoot from the homogeneous roughness curve, $e_{i,h}$ would be zero. A positive value of $e_{i,h}$ means that on average, there is an overshoot while a negative value of $e_{i,h}$ means that there is an undershoot relative to the corresponding homogeneous curves. The values of $e_{i,h}$ are tabulated in **Table 2**.

Table 2 shows cases with decreasing magnitude of overshoot in the following order: PQR > PRQ > QPR and cases with decreasing magnitude of undershoot in the following order: RQP > RPQ > QRP. A consistent trend is observed in all the cases with different plate lengths. The most significant averaged overshoot (1.87%) and undershoot (1.14%) are observed (PQR and RQP, respectively).

5.2 Overall skin friction C_F

Following the local skin friction analysis from the previous subsection, it is desirable to estimate the overall skin friction coefficient C_F of the plates. This allows us to see the influence of individual roughness height k_s or the combination of it in a more global way. The overall skin friction coefficient C_F is given in Eq. (9). It is related to the local skin friction coefficient $c_f(x)$ by the relation described in Eq. (23).

$$C_F = \frac{\int_0^L c_f(x) \, dx}{L} \qquad (23)$$

The corresponding overall skin friction coefficients C_{F_s} for the plate segments 1, 2, and 3 are given, respectively, as described in Eq. (24). Where x is the distance in the streamwise direction with the origin at the leading edge of the plate. The lengths of plate segments 1, 2, and 3 are denoted as L_1, L_2 and L_3, respectively.

Table 2 summarizes the overall C_F and those for each plate segments ($C_{F;1}$, $C_{F;2}$, $C_{F;3}$). **Table 2** shows that for the homogeneous cases, both the smooth SSS and the three rough cases (PPP, QQQ, and RRR), the overall friction coefficient C_F decreases as the flow move from the upstream to downstream ($C_{F;1} > C_{F;2} > C_{F;3}$). The RRR case has the largest C_F among the three homogeneous roughness cases due to its highest k_s value.

5.2.1 Quantification of the overall skin friction between rough surface and smooth surface

Having obtained the overall skin friction C_F from individual plates, quantification the change in drag penalty between one case to another is made in a more simplified way. The first analysis we are interested in quantifies the roughness wall effects (both homogenous and inhomogeneous cases) on the overall skin friction relative to the smooth wall case. A percent increase in overall skin friction $e_{r;s}$ due to roughness effects is defined as described in Eq. (24). The subscripts r and s refer to rough and smooth, respectively. The results are tabulated in **Table 2**.

$$e_{r;s} = \frac{C_{Fr} - C_{Fs}}{C_{Fs}} \times 100\% \tag{24}$$

Table 2 shows that the RRR case results in the highest $e_{r;s}$ due to the highest k_s with a value of 75.23%. For the same reasons, the smallest $e_{r;s}$ resulted from the H240_PPP case, with a value of 18.65%. It is interesting to note that the homogeneous H120_QQQ case with $k_s = 325\ \mu m$ experienced an increase in drag penalty of 52.62% compared to the smooth wall case. Such a value of roughness height represents heavy slime [16] or fouled with light calcareous tube-worm fouling [51].

A similar rough and smooth wall $e_{r;s}$ analysis is also conducted for the inhomogeneous cases. The results show that the homogenous QQQ case (with $k_s = 325\ \mu m$ has a higher percent increase in overall skin friction $e_{r;s}$ than the inhomogeneous cases. Although the averaged roughness heights for the inhomogeneous cases are the same as the QQQ roughness height, their representative roughness heights are smaller than the QQQ roughness height and depend on the sequence roughness heights in the streamwise direction. It is observed that the values of $e_{r;s}$ for the inhomogeneous cases increase monotonically in the following order: PQR < PRQ < QPR < QRP < RPQ < RQP.

5.2.2 Quantification of the overall skin friction between inhomogeneous and homogeneous rough surface

Apart from looking at the percent increase in overall skin friction $e_{r;s}$ between the rough wall and smooth wall, it is also desirable to quantify the effects of roughness inhomogeneity (combination of PQR) on the overall skin friction C_F with respect to the homogeneous QQQ baseline case. For that purpose, a percent decrease between an inhomogeneous roughness case and the homogeneous QQQ case is defined in Eq. (25).

$$e_{i;q} = \frac{C_{F;i} - C_{F;QQQ}}{C_{F;QQQ}} \times 100\% \tag{25}$$

The subscript i refers to inhomogeneous roughness (variation of PQR) while the subscript QQQ refers to the homogenous rough wall base case. As has been noted above, the friction coefficient $C_{F;QQQ}$ is chosen as a reference because the arithmetic average of k_s for the inhomogeneous cases is equal to that of the homogeneous QQQ case. Calculating skin friction from measured surface roughness would normally use a single roughness value which generally comes from the average of measurements over the hull. Thus, $e_{i;q}$ represents the error of assuming a single (average) roughness value for an in-homogeneously rough hull. A negative value of $e_{i;q}$ indicates that the C_F values of the inhomogeneous cases (combination of PQR) are lower than the homogeneous base case (QQQ). The opposite is valid for a positive value of $e_{i;q}$. **Table 2** shows that the $e_{i;q}$ values are negative for all the inhomogeneous cases, indicating that the C_F values of all the inhomogeneous cases (combination of PQR) are lower than the homogeneous base case (QQQ). For the same plate length, the magnitude of $e_{i;q}$ decreases orderly in the sequence of PQR to RQP, indicating that the order of roughness arrangement plays a key role.

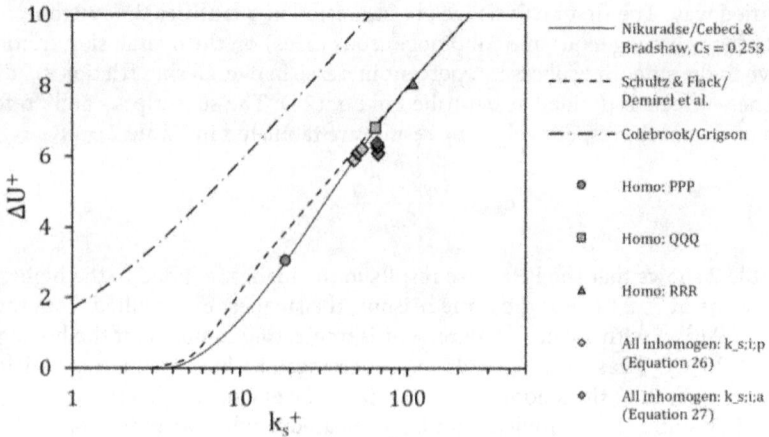

Figure 6.
Plotting the results of the verification of the predicted k_s values for inhomogenous cases with $k_{s;i;p}$ and $k_{s;i;a}$.

5.2.3 Prediction of a representative roughness height for an inhomogeneous rough surface

Having discussed the local and overall skin frictions, a question may arise: "Can we predict a representative roughness height k_s for an inhomogeneous rough surface?" We do this prediction with the help of the method from Granville [45], namely Eqs. 18 and 19, as well as the minimum error optimization help iterated by the solver to calculate the portion of each segment, L_1, L_2, and L_3. The result of this work creates Eq. (26), where it is found that the segment L_1 has a more significant portion, namely 29.7%. The L_2 segment has a portion of 24.3% and L3 $\approx$ 23.1%, where if all the portion of segments are totaled, the value will be smaller than the average value of the inhomogeneous k_s (see Eq. (27)), $k_{s;i;p} < k_{s;i;a}$.

$$k_{s;i;p} \approx 29.7\% \cdot k_{s;1} + 24.3\% \cdot k_{s;2} + 23.1\% \cdot k_{s;3} \qquad (26)$$

$$k_{s;i;a} = \frac{k_{s;1} + k_{s;2} + k_{s;3}}{3} \qquad (27)$$

The verification of the equation for predicting the k_s value for the inhomogeneous case is done with the aid of the Granville method [45], which plots the ΔU^+ and k_s^+, as explained in Eqs. (18 and 19), respectively. The results of the verification are plotted in **Figure 6**, where not only the results of the calculation $k_{s;i;p}$, but also the results of the calculation of $k_{s;i;a}$ are plotted. From the plot results, it can be seen that, prediction using $k_{s;i;p}$ got a very good match to the roughness function used in this simulation, namely from Cebeci and Bradshaw [39].

6. Conclusions

Rough-wall turbulent boundary layer flow is a complex physical phenomenon that increases skin friction drag compared to the smooth wall case. Because the surface roughness of a ship hull fouled with biofoulings or other types of hull imperfections is often found to be inhomogeneous, it is crucial to consider inhomogeneous roughness in addition to homogeneous roughness. In this study, the effects of roughness inhomogeneity on the skin friction drag are investigated by modeling the inhomogeneous roughness pattern in a simplified manner using step changes in the equivalent sand grain roughness heights k_s, denoted as P, Q, and R.

Some of the findings from this study include:

- Such a step-change in the roughness height results in overshoot/undershoot of the local skin friction coefficient c_f, followed by a relaxation where the c_f value is slowly returning to the underlying rough wall homogeneous c_f value. A step up in k_s results in an overshoot, while a step down in k_s results in an undershoot in the c_f values. In some cases where the jump happens over two significantly different k_s values (i.e., P to R), the relaxation rate is very slow and unable to recover over the given streamwise distance fully. Cases with decreasing magnitude of overshoot are found in the following order: PQR > PRQ > QPR, and cases with decreasing magnitude of undershooting are in the following order: RQP > RPQ > QRP.

- The sequence of roughness arrangement in a streamwise inhomogeneous roughness pattern plays a key role in the resulting overall skin friction coefficient C_F. It is found that C_F increases in the following order: PQR < PRQ < QPR < QRP < RPQ < RQP. This result is further reflected in the predicted k_s values, showing k_s increases in the same order: PQR < PRQ < QPR < QRP < RPQ < RQP. A change of roughness near the leading edge of the plate has a much more significant effect on the overall skin friction coefficient C_F than a change of this near the trailing edge. In practical terms, limiting (cleaning) the biofouling from the bow of a ship is of greatest benefit and should be prioritized.

- The overall skin friction (C_F) in inhomogeneous cases (e.g. PQR) is smaller than the k_s value with the same mean. Thus, the mean value of k_s for the inhomogeneous case is less suitable for predicting the C_F value. A new way to predict the k_s value for the inhomogeneous case has been proposed in this paper (See Eq. (26)), where the value is very close to the simulation results that have been carried out.

Acknowledgments

This research project was supported by the Ministry of Research, Technology, and National Innovation and Research Agency (Kemenristek – BRIN) of the Republic of Indonesia under the World Class Professor 2019 Grant (Contract No. T/42/D2.3/KK.04.05/2019) and Master to Doctorate Program for Excellent Graduate (PMDSU) scholarship program batch III (Contract No. 1277/PKS/ITS/2020).

Conflict of interest

The authors declare that they have no known competing financial interests or personal relationships that could have appeared to influence the work reported in this paper.

Author details

I. Ketut Aria Pria Utama*, I. Ketut Suastika and Muhammad Luqman Hakim
Department of Naval Architecture, Institut Teknologi Sepuluh Nopember,
Surabaya, Indonesia

*Address all correspondence to: kutama@na.its.ac.id

IntechOpen

References

[1] Hasan MI, Khafeef MJ, Mohammadi O, Bhattacharyya S, Issakhov A. Investigation of Counterflow Microchannel Heat Exchanger with Hybrid Nanoparticles and PCM Suspension as a Coolant. Menni Y, editor. Mathematical Problems in Engineering. 2021 Mar 23; 2021:1–12.

[2] Soni MK, Tamar N, Bhattacharyya S. Numerical simulation and parametric analysis of latent heat thermal energy storage system. Journal of Thermal Analysis and Calorimetry. 2020 Sep 4; 141(6):2511–26.

[3] Bhattacharyya S, Chattopadhyay H, Biswas R, Ewim DRE, Huan Z. Influence of Inlet Turbulence Intensity on Transport Phenomenon of Modified Diamond Cylinder: A Numerical Study. Arabian Journal for Science and Engineering. 2020 Feb 5;45(2):1051–8.

[4] Kumar S, Kumar R, Goel V, Bhattacharyya S, Issakhov A. Exergetic performance estimation for roughened triangular duct used in solar air heaters. Journal of Thermal Analysis and Calorimetry . 2021 May 24;

[5] Murmu SC, Bhattacharyya S, Chattopadhyay H, Biswas R. Analysis of heat transfer around bluff bodies with variable inlet turbulent intensity: A numerical simulation. International Communications in Heat and Mass Transfer . 2020 Oct;117:104779.

[6] Alam MW, Bhattacharyya S, Souayeh B, Dey K, Hammami F, Rahimi-Gorji M, et al. CPU heat sink cooling by triangular shape micro-pin-fin: Numerical study. International Communications in Heat and Mass Transfer . 2020 Mar;112:104455.

[7] Paul AR, Bhattacharyya S. Analysis and Design for Hydraulic Pipeline Carrying Capsule Train. Journal of

Pipeline Systems Engineering and Practice . 2021 May;12(2):04021003.

[8] Suastika K, Hidayat A, Riyadi S. Effects of the Application of a Stern Foil on Ship Resistance: A Case Study of an Orela Crew Boat. International Journal of Technology . 2017 Dec 26;8(7):1266.

[9] Smith TWP, Jalkanen JP, Anderson BA, Corbett JJ, Faber J, et al. Third IMO Greenhouse Gas Study 2014. International Maritime Organization (IMO). 2014;

[10] Buhaug Ø, Corbett J., Endresen Ø, Eyring V, Faber J, Hanayama S, et al. Second IMO GHG Study2009. International Maritime Organization (IMO). 2009;

[11] Molland AF, Turnock SR, Hudson DA, Utama IKAP. Reducing ship emissions: A review of potential practical improvements in the propulsive efficiency of future ships. Transactions of the Royal Institution of Naval Architects Part A: International Journal of Maritime Engineering. 2014; 156(PART A2):175–88.

[12] Wang H, Lutsey N. Long-term potential to reduce emissions from international shipping by adoption of best energy-efficiency practices. Transportation Research Record. 2014; 2426:1–10.

[13] Kodama Y, Kakugawa A, Takahashi T, Kawashima H. Experimental study on microbubbles and their applicability to ships for skin friction reduction. International Journal of Heat and Fluid Flow. 2000;21 (5):582–8.

[14] Hakim ML, Nugroho B, Chin RC, Putranto T, Suastika IK, Utama IKAP. Drag penalty causing from the roughness of recently cleaned and

painted ship hull using RANS CFD. CFD Letters. 2020;12(3):78–88.

[15] Monty JP, Dogan E, Hanson R, Scardino AJ, Ganapathisubramani B, Hutchins N. An assessment of the ship drag penalty arising from light calcareous tubeworm fouling. Biofouling. 2016;32(4):451–64.

[16] Schultz MP. Effects of coating roughness and biofouling on ship resistance and powering. Biofouling. 2007;23(5):331–41.

[17] Hakim ML, Nugroho B, Nurrohman MN, Suastika IK, Utama IKAP. Investigation of fuel consumption on an operating ship due to biofouling growth and quality of anti-fouling coating. IOP Conference Series: Earth and Environmental Science. 2019; 339(1):012037.

[18] Schultz MP, Bendick JA, Holm ER, Hertel WM. Economic impact of biofouling on a naval surface ship. Biofouling. 2011;27(1):87–98.

[19] Nikuradse J. Laws of flow in rough pipes [English translation of Stromungsgesetze in rauhen Rohren]. VDI-Forschungsheft 361 Beilage zu "Forschung auf dem Gebiete des Ingenieurwesens" [Translation from NACA Technical Memorandum 1292]. 1933;

[20] Hama F. Boundary-layer characteristics for smooth and rough surfaces. Transactions - The Society of Naval Architects and Marine Engineers. 1954;62:333–58.

[21] Demirel YK, Turan O, Incecik A. Predicting the effect of biofouling on ship resistance using CFD. Applied Ocean Research. 2017;62:100–18.

[22] Song S, Demirel YK, Atlar M. An investigation into the effect of biofouling on the ship hydrodynamic characteristics using CFD. Ocean Engineering . 2019;175:122–37.

[23] Andersson J, Oliveira DR, Yeginbayeva I, Leer-Andersen M, Bensow RE. Review and comparison of methods to model ship hull roughness. Applied Ocean Research. 2020;99: 102119.

[24] Song S, Demirel YK, Atlar M. Penalty of hull and propeller fouling on ship self-propulsion performance. Applied Ocean Research. 2020;94: 102006.

[25] Song S, Shi W, Demirel YK, Atlar M. The effect of biofouling on the tidal turbine performance. Applied Energy Symposium: MIT A+B. 2019.

[26] Atencio BN, Chernoray V. A resolved RANS CFD approach for drag characterization of antifouling paints. Ocean Engineering. 2019;171:519–32.

[27] Demirel YK, Khorasanchi M, Turan O, Incecik A, Schultz MP. A CFD model for the frictional resistance prediction of antifouling coatings. Ocean Engineering. 2014;89:21–31.

[28] Schultz MP. Frictional Resistance of Antifouling Coating Systems. Journal of Fluids Engineering. 2004;126(6):1039–47.

[29] Song S, Demirel YK, Atlar M, Dai S, Day S, Turan O. Validation of the CFD approach for modelling roughness effect on ship resistance. Ocean Engineering. 2020;200:107029.

[30] Alamsyah MA, Hakim ML, Utama IKAP. Study of Shear and Pressure Flow on the Variation of Ship Hull Shapes as One of the Biofouling Growth Factors. In: Proceedings of the 3rd International Conference on Marine Technology. SCITEPRESS - Science and Technology Publications; 2018. p. 97–105.

[31] Antonia RA, Luxton RE. The response of a turbulent boundary layer to a step change in surface roughness

Part 1. Smooth to rough. Journal of Fluid Mechanics. 1971;48(4):721–61.

[32] Antonia RA, Luxton RE. The response of a turbulent boundary layer to a step change in surface roughness. Part 2. Rough-to-smooth. Journal of Fluid Mechanics. 1972;53(4):737–57.

[33] Chung D, Hutchins N, Schultz MP, Flack KA. Predicting the Drag of Rough Surfaces. Vol. 53, Annual Review of Fluid Mechanics. 2021.

[34] Pendergrass W, Arya SPS. Dispersion in neutral boundary layer over a step change in surface roughness —I. Mean flow and turbulence structure. Atmospheric Environment (1967). 1984;18(7):1267–79.

[35] Suastika IK, Hakim ML, Nugroho B, Nasirudin A, Utama IKAP, Monty JP, et al. Characteristics of drag due to streamwise inhomogeneous roughness. Ocean Engineering. 2021;223: 108632.

[36] Song S, Demirel YK, Muscat-Fenech CDM, Sant T, Villa D, Tezdogan T, et al. Investigating the effect of heterogeneous hull roughness on ship resistance using cfd. Journal of Marine Science and Engineering. 2021;9(2).

[37] Ferziger JH, Perić M. Computational Methods for Fluid Dynamics. Computational Methods for Fluid Dynamics. 2002.

[38] Shih T-H, Liou WW, Shabbir A, Yang Z, Zhu J. A new k-ε eddy viscosity model for high reynolds number turbulent flows. Computers & Fluids. 1995;24(3):227–38.

[39] Cebeci T, Bradshaw P. Momentum transfer in boundary layers. New York: Hemisphere Publishing Corporation; 1977.

[40] Mitchell R, Webb M, Roetzel J, Lu F, Dutton J. A study of the Base Pressure Distribution of a Slender Body of Square Cross-Section. In: 46th AIAA Aerospace Sciences Meeting and Exhibit. Reston, Virigina: American Institute of Aeronautics and Astronautics; 2008.

[41] Anderson J. Computational Fluid Dynamics: The Basics with Applications. 1995. McGrawhill Inc. 1995;

[42] Cant S. S. B. Pope, Turbulent Flows, Cambridge University Press, Cambridge, U.K., 2000, 771 pp. Combustion and Flame. 2001;125(4).

[43] Granville P. The Frictional Resistance and Turbulent Boundary Layer of Rough Surfaces. Journal of Ship Research. 1958;2(04):52–74.

[44] Schoenherr KE. Resistance of flat surfaces. Trans SNAME. 1932;40:40: 279-313.

[45] Granville PS. Three Indirect Methods for the Drag Characterization of Arbitrarily Rough Surfaces on Flat Plates. Journal of Ship Research. 1987;31 (1):70–7.

[46] Flack KA, Schultz MP, Rose WB. The onset of roughness effects in the transitionally rough regime. International Journal of Heat and Fluid Flow. 2012;35:160–7.

[47] Schultz MP, Flack KA. The rough-wall turbulent boundary layer from the hydraulically smooth to the fully rough regime. Journal of Fluid Mechanics. 2007;580:381–405.

[48] Moody LF. Friction factors for pipe flow. Transaction of the ASME. 1944;66: 671–84.

[49] Demirel YK, Uzun D, Zhang Y, Fang HC, Day AH, Turan O. Effect of barnacle fouling on ship resistance and powering. Biofouling. 2017;33(10):819–34.

[50] Andreopoulos J, Wood DH. The response of a turbulent boundary layer

to a short length of surface roughness.
Journal of Fluid Mechanics. 1982;118
(1):143.

[51] Monty JP, Dogan E, Hanson R,
Scardino AJ, Ganapathisubramani B,
Hutchins N. An assessment of the ship
drag penalty arising from light
calcareous tubeworm fouling.
Biofouling. 2016;32(4):451–64.

Chapter 8

CFD Combustion Simulations and Experiments on the Blended Biodiesel Two-Phase Engine Flows

Vinay Atgur, Gowda Manavendra,
Gururaj Pandurangarao Desai
and Boggarapu Nageswara Rao

Abstract

Biodiesels are the promising sources of alternative energy. Combustion phenomenon of blended biodiesels differs to those of diesel due to changes in physio-chemical properties. Experimental investigations are costly and time-consuming process, whereas mathematical modeling of the reactive flows is involved. This chapter deals with combustion simulations on four-stroke single-cylinder direct injection compression ignition engine running at a constant speed of 1500 rpm, injection timing of 25° BTDC with diesel and 20% blend of Jatropha biodiesel. Standard finite volume method of computational fluid dynamics (CFD) is capable of simulating two-phase engine flows by solving three-dimensional Navier–Stokes equations with k-ε turbulence model. Combustion simulations have been carried out for half-cycle by considering the two strokes compression and expansion at zero load condition. The model mesh consists of 557,558 elements with 526,808 nodes. Fuel injection begins at 725° and continues till 748° of the crank angle. Charge motion within the cylinder, turbulent kinetic energy, peak pressure, penetration length, and apparent heat release rate are analyzed with respect to the crank angle for diesel and its B-20 Jatropha blend. Experimental data supports the simulation results. B-20 Jatropha blend possesses similar characteristics of diesel and serves as an alternative to diesel.

Keywords: combustion, simulation, crank angle, compression, fuel spray

1. Introduction

Combustion plays a vital role in the chemical, electrical and transportation industries [1–3].

Understanding of the combustion phenomenon through experimentation is involved and expensive. In such situations, numerical simulations act as an alternative platform. There is a need to develop mathematical models for the reactive flows. Due to variation of density, the heat transfer and fluid motion inside the engine is unsteady and turbulent. Most of the simplified real process versions are based on the idealization of the cylindrical geometry models. Combustion phenomenon is greatly influenced by fuel properties, fuel preparation and fuel distribution inside the cylinder. The advancement in computer technology is helpful in solving the complicated equations relevant to the turbulence-chemistry interactions.

Jafarmadar and Zehni [4] have studied the high-speed diesel engine combustion using AVL-FIRE code CFD. They have analyzed the peak pressure rise and heat release rate. They have made comparison of numerical simulations with experiments varying the fuel injection pressure. The KIVA group of codes will be helpful in performing diesel engine simulations with less computational time. The enhanced code and coarse meshes are utilized to simulate combustion in a heavy-duty Mitsubishi Heavy Industries diesel engine for the service loads, speeds, and injection pressure. The normal simulation time from IVC to EVO is reduced from 60 hours to 1 hour using 12 processors [5]. Various researchers have developed alternative codes and models for minimizing the simulation time of combustion [6–8].

Mirko Baratta et al. [9] have utilized CFD models and analyzed laminar flame speed for different fuel composition and mixture dilution rates. Michela Costa et al. [10] have performed simulations on premixed syngas and biodiesel as pilot injection. The combustion efficiency decreases, exhaust gas temperature and thermal efficiency increase with increasing the % of syngas. The reduced chemical kinetics model gives an improved solution. Amin Maghbouli et al. [11] have used a 3D-CFD/ Chemical kinetics framework model to investigate the diesel engine/gas dual-fuel engine combustion process. Methane and n-heptane are used as natural gas representatives. Source terms in conservation equations of energy and species are calculated by integrating CHEMKIN solver into KIVA code. Pressure, igniton delay and heat release rate are in good agreement with experiments. Vijayshree and Ganesan [12] have performed CFD simulations for designing IC engine through combustion process analysis.

CFD studies thus provide flow visualization, optimal engine parameters and knowledge in combustion phenomenon, which are difficult to acquire from experiments. Experimental investigations are involved in obtaining the penetration length, velocity distribution, swirl ratio, tumble ratio and heat release rate. Modifications in engine design and parameters are difficult to implement. The task will be definitely a time-consuming process. CFD serves as a versatile and powerful tool for designing IC engine and gives insight into the complex fluid dynamics. Experiments on various blends (5–30%) indicate B-20 blend as viable in terms of performance and emission. Combustion simulations help in minimizing engine bench tests.

Comparative studies are made in this article to examine the combustion behaviour of diesel and B-20 blend of Jatropha. Combustion simulations have been performed utilizing ANSYS Fluent 15.0 version. Combustion simulations are in line with those of DSC (differential scanning calorimetry) analysis with a heating rate of 10°C/min in atmospheric air.

2. Modeling and simulations

This section deals with the combustion simulations on the four-stroke single-cylinder direct injection compression ignition engine running at a constant speed of 1500 rpm, injection timing of 25° BTDC with diesel and 20% blend of Jatropha biodiesel. Standard finite volume method of computational fluid dynamics (CFD) is capable of simulating the two-phase engine flow. Three-dimensional Navier–Stokes equations are solved with k-ε turbulence model. The details of combustion simulations carried out for half-cycle by considering the two strokes compression and expansion at zero load condition are presented below. **Figure 1** shows the front view and 3D view of the geometry model. The model is a four-stroke diesel engine with 4 valves and 4 ports for air sucking inlets and hot gas outlets. The green portion in **Figure 1** is the assembly of a cylinder and piston. Specifications are made from the standard engine KIRLOSKAR AV-1 model. The mathematical models in CFD begin with representation of combustion chamber geometry (meshing of engine). The meshing of the geometry model **Figure 2** is generated using the pre-processor of ANSYS Fluent 15.0 version.

The fluid chamber bottom of ICE is modelled with 450,570 elements and 470,654 nodes. The fluid chamber top of ICE is modelled with 9795 elements and 13,544 nodes. Fluid piston of ICE is modelled with 66,443 elements and 73,360 nodes. The model domain consists of 526,808 elements and 557,558 nodes. Mesh

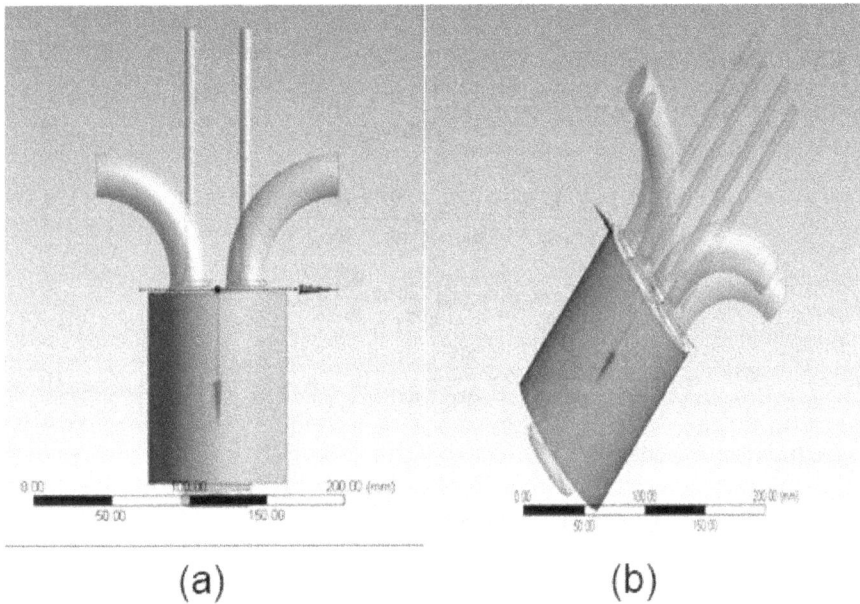

(a)   (b)

Figure 1.
Geometry model. (a) Front view. (b) 3D view.

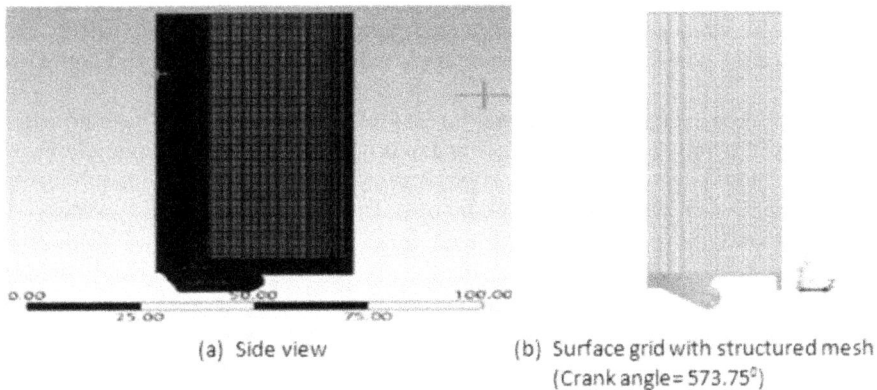

(a) Side view   (b) Surface grid with structured mesh
                       (Crank angle = 573.75°)

Figure 2.
Geometry mesh.

parameters of IC sector are: Reference size = 0.947 mm; Minimum mesh size = 0.19 mm; Maximum mesh size = 0.474 mm; and the chamber body mesh size = 1.487 mm with 3 inflation layers.

The complex physical phenomenon of combustion flows in IC engines (see **Figure 2**) can be understood by solving the following 3-Dimensional Naiver-Stokes (N-S) equations with the Reynold's Average Navier–Stokes (RANS) model [13–15] and the k-ε turbulence model [16, 17].

N-S equations:

$$\frac{\partial v_j}{\partial x_j} = 0 \tag{1}$$

$$\rho\left(\frac{\partial v_i}{\partial t} + v_j\frac{\partial v_i}{\partial x_j}\right) = -\frac{\partial P}{\partial x_i} + \mu\frac{\partial}{\partial x_j}\left(\frac{\partial v_i}{\partial x_j}\right) \tag{2}$$

RANS model:

$$\frac{\partial \bar{v}_i}{\partial x_j} = 0 \tag{3}$$

$$\rho\left(\frac{\partial \bar{v}_i}{\partial t} + \bar{v}_j\frac{\partial \bar{v}_i}{\partial x_j}\right) = -\frac{\partial \bar{P}}{\partial x_i} + \mu\frac{\partial}{\partial x_j}\left(\frac{\partial \bar{v}_i}{\partial x_j}\right) - \rho\frac{\partial \overline{u'_i u'_j}}{\partial x_j} \tag{4}$$

$$-\rho\overline{u'_i u'_j} = \mu_t\left(\frac{\partial \bar{v}_i}{\partial x_j} + \frac{\partial \bar{v}_j}{\partial x_i}\right) \tag{5}$$

$$\rho\left(\frac{\partial \bar{v}_i}{\partial t} + \frac{\partial \bar{v}_i}{\partial x_j}\right) = -\frac{\partial \bar{P}}{\partial x_i} + \frac{\partial}{\partial x_j}\left[(\mu + \mu_t)\left(\frac{\partial \bar{v}_i}{\partial x_j}\right)\right] \tag{6}$$

k-ε turbulence model:

$$\rho\left(\frac{\partial k}{\partial t} + \frac{\partial(kv_j)}{\partial x_j}\right) = \frac{\partial}{\partial x_j}\left[\left(\mu + \frac{\mu_t}{\sigma_k}\right)\frac{\partial k}{\partial x_j}\right] + \mu_t\gamma^2 - \rho\xi \tag{7}$$

Here ρ is density; μ is dynamic viscosity; u and v are velocity components in x and y directions; p refers to pressure; μ_t is the eddy or turbulent viscosity; ξ is the rate of heat dissipation; k is the turbulent kinetic energy; $\sigma_k \approx 1$, is model constant. The RANS models are created with variations referred to over bars and apostrophes in Eqs. (3)–(6) for incompressible fluids. Eqs. (5) and (6) are used for evaluating the Reynolds 'stress equivalent to the average gradients of velocity (equivalent to shear stress). The k-ε turbulence model (7) will be used for the gas and liquid phase [18].

Finite volume equations are crucial for CFD simulations to handle fluid boundary layers on surfaces [19, 20]. The shear gap is heavily influenced by boundary conditions. Fluid near the wall (i.e., layer close to the wall) is the viscous sub-layer, whereas above this layer is dominated by turbulent shears. The laminar pressure-stress relation is

$$\frac{v C_\mu^{0.25} k^{0.25}}{\frac{\tau_w}{\rho}} = v^* = y^* = \frac{\rho C_\mu^{0.25} k^{0.25} y}{\mu} \tag{8}$$

Here y is the cell's close-to-wall position; v is the medium fluid speed; and C_μ is nearly equal to 0.09. It is more precise to calculate near-wall gradients using constrained equivalences when the laminar stress–strain relationship is important.

The near-wall grid involves substantially larger number of cells, which increases computational time.

The temperature at the ICE-cylindrical chamber bottom and top surfaces is 567 K. The ICE-cylindrical-piston wall temperature is also specified as 567 K. The wall temperature of ICE-piston is 645 K. The wall temperature is 602 K on ICE-sector-top-faces. Relaxation of crank angles are: Engine speed = 1500 rev/min.; Crank radius = 55 mm; Piston pin-offset = 0 mm; Connecting rod length = 165 mm; Cylinder bore length = 110 mm; and Cylinder bore diameter = 90 mm. **Table 1**

Parameter	Dimension
X-Position	0
Y-Position	−0.00012
Z-Position	2E-05
X-Axis	0
Y-Axis	−0.34202
Z-Axis	0.939693
Diameter (m)	2.54E-4
Evaporating Species	C_7h_{16}
Temperature (K)	366.7
Start Crank Angle (deg)	721
End Crank Angle (deg)	742.5
Cone Angles (deg)	9
Cone Radius (deg)	1.27E-4
Total Flow Rate (kg/s)	1.3333E-05
Velocity Magnitude (m/s)	468

Table 1.
Injection Properties.

Fuel	Calorific value (MJ/kg)	Kinematic Viscosity @40°C (cst)	Cetane value	Density (kg/m^3)	Flash point (°C)	Pour point (°C)	Cloud point (°C)
DIESEL	44.22	2.87	47.8	840	76	−3	6.5
JOME	39.79	4.73	52	862.2	182.5	3	3
B-20JOME	44.10	3.99	49	840.2	93.5	−3	4

Table 2.
Biodiesel and diesel properties [2].

Fuel	Thermal Conductivity (W/mk)	Specific Heat (MJ/m^{3}K)	Thermal Diffusivity (mm^2/s)
Diesel	0.3390	0.2563	1.323
JOME	0.2537	0.5773	0.4083
B-20 JOME	0.3090	0.3894	0.7934

Table 3.
Thermo-physical properties of diesel and biodiesel.

provides the injection properties. **Tables 2** and **3** provide thermo-physical properties of diesel, Jatropha oil methyl ester (JOME) and its B-20 blend for combustion simulations. The properties are measured from TPS 500S with Kapton and Teflon sensors.

Fuel starts to penetrate into the combustion chamber at 728°C for the diesel as well as biodiesel. Due to the high viscosity of the B-20 Jatropha leads to poor atomization. As in [21], the temperature rise during the fuel spray is around 2770°C for diesel and 2670°C for biodiesel (see **Figures 3** and **4**). As in [22], B-20 Jatropha exhibits high magnitude of velocity for atomization due to viscosity on fuel spray. Hot air presence prior to the fuel injection evaporates the fuel just beyond fixed length (which is called a break-up the length). Engine cylinder spray is around 50–100 atmosphere. At that time, fuel is injected into the chamber. Since the

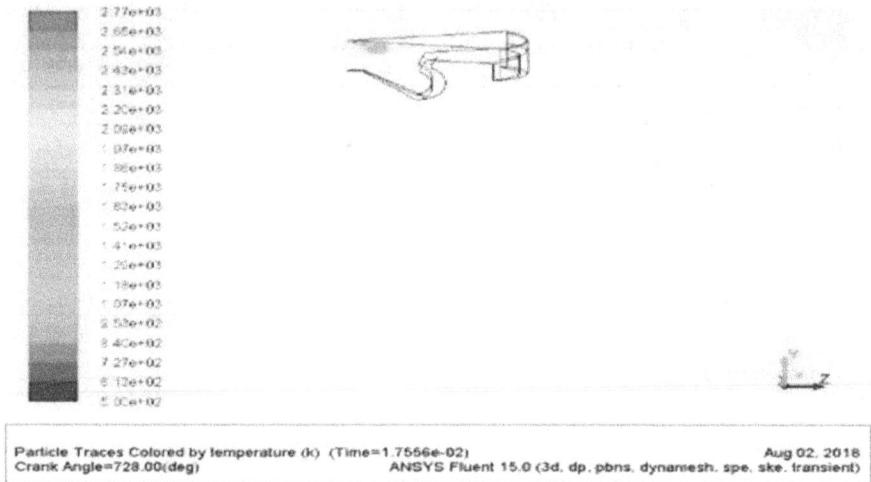

Particle Traces Colored by temperature (k) (Time=1.7556e-02) Aug 02, 2018
Crank Angle=728.00(deg) ANSYS Fluent 15.0 (3d, dp, pbns, dynamesh, spe, ske, transient)

Figure 3.
Visualization of spray at 728° for diesel.

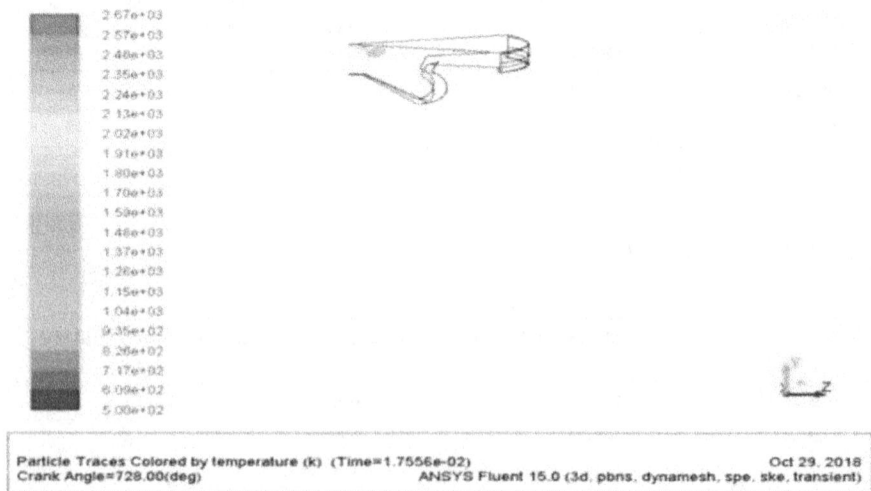

Particle Traces Colored by temperature (k) (Time=1.7556e-02) Oct 29, 2018
Crank Angle=728.00(deg) ANSYS Fluent 15.0 (3d, pbns, dynamesh, spe, ske, transient)

Figure 4.
Visualization of spray at 728° for B-20 Jatropha.

Contours of Velocity Magnitude (m/s) (Time=1.8444e-02)   Aug 02, 2018
Crank Angle=736.00(deg)   ANSYS Fluent 15.0 (3d, dp, pbns, dynamesh, spe, ske, transient)

Figure 5.
Velocity contour plot at the time of spray for diesel.

Contours of Velocity Magnitude (m/s) (Time=1.8445e-02)   Oct 29, 2018
Crank Angle=736.00(deg)   ANSYS Fluent 15.0 (3d, pbns, dynamesh, spe, ske, transient)

Figure 6.
Velocity contour plot at the time of spray for B-20 Jatropha.

high-velocity jet has to mix with compressed air in a small interval of time thereby B-20 blend exhibits slightly low velocity magnitude (see **Figures 5** and **6**).

Fuel injection starts at 724°C and ends at 740°C. During fuel injection temperature varies from 500 to 2770°C. But at the end of the compression stroke, diesel temperature varies from 500 to 2360°C, whereas biodiesel temperature varies from 500 to 2180°C. Since B-20 blend is having less heat of vaporization when compared to that of diesel, heat transfer lowers the local air temperature as observed in [23]. Similarly, the magnitude of velocity for B-20 blend is slightly lower than that of diesel (see **Figures 5** and **6**).

Figures 7 and **8** show the variation of temperature after the combustion for diesel and B-20 Jatropha. Temperature varies from 443 to 705°C for the diesel, whereas it varies from 437 to 685°C for biodiesel. Slightly low temperature variation is noticed for the B-20 blend. This could be due to high diffusion burning phase for

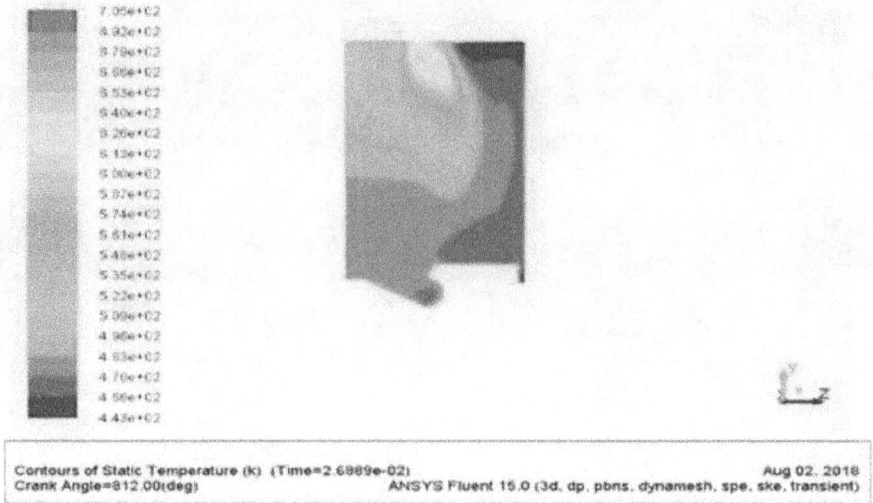

Figure 7.
Temperature distribution after combustion for diesel.

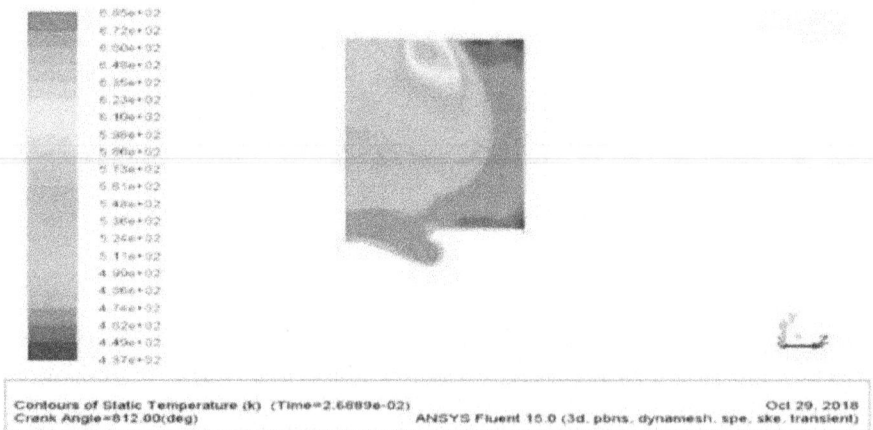

Figure 8.
Temperature distribution after combustion for B-20 Jatropha.

the biodiesel. Due to increase in volatility of the slow-burning biodiesel, burning time is significantly high for B-20 as in [24].

It is noted at the end of the expansion stroke that the velocity magnitude for diesel varies up to 9.34 m/s, whereas in case of B-20 blend, it varies up to 9.4 m/s. Some amount of residual gases present in the engine at the end of expansion stroke.

3. Differential Scanning Calorimetry (DSC) Experiments

DSC experiments are performed under air atmosphere to diesel, Jatropha oil methyl ester (JOME), and its B-20 blend at 10°C/min in Universal TA instruments with alumina pan. Performing experiments on liquid samples is difficult due to evaporation and non-stability of complex during heating. Hermetically sealed pans

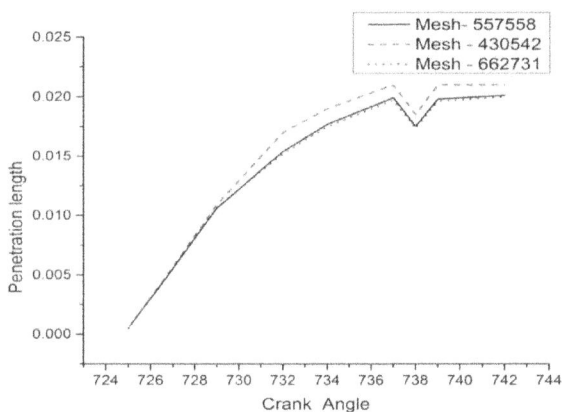

Figure 9.
Mesh Convergence study.

are used in experiments with universal cramper. Two types of calibration are performed on instrument: (i) Initially with T-Zero (Temperature) calibration; and (ii) Enthalpy calibration. Initially Experiments are performed without the samples to get baseline further with known material (sapphire or Al_2O_3) heat of fusion is calculated by heating up to its melting point. Heat of fusion value is compared with theoretical estimates.

4. Mesh convergence study

Convergence study is made by varying mesh (See **Figure 9**). Computational domain is chosen for three different meshes (430,542, 557,558 and 662,731). Maximum deviation of 9.4% is observed in results by increasing the number of meshes from 430,542 to 557,558. Further increasing to 662,731, maximum deviation of 1.3% is observed. From this study, number of meshes finalized for computation is 557,558.

5. Results and discussion

Deformation on the working fluid increases due to viscous shear stress. Thereby its internal energy increases at the expense of its turbulent kinetic energy. During compression, the airflow is forced into the piston and the swirl rotational velocity increases at the end of the compression stroke. The radius is reduced while the momentum is conserved leading to increase in angular velocity. When the piston moves down, reverse trend happens. The flow slows down due to the friction against the combustion chamber walls [25].

Velocity magnitude in **Figure 10** shows little variation from 570° to 725° and large variation where injection starts at 725° and ends at 748°. Biodiesels show slightly low velocity magnitudes resulting in the time delay. The penetration length in **Figure 11** can be divided into 3 phases as in. [26]. In the initial phase, there is no penetration length (i.e., zero) for the crank angles from 722° to 725°. In the second phase, the penetration length increases rapidly from starting to ending of injection period for the crank angles from 725° to 737°, which indicates more amount of fuel injected inside the engine cylinder. Similar phenomenon is observed for the biodiesel with less penetration length.

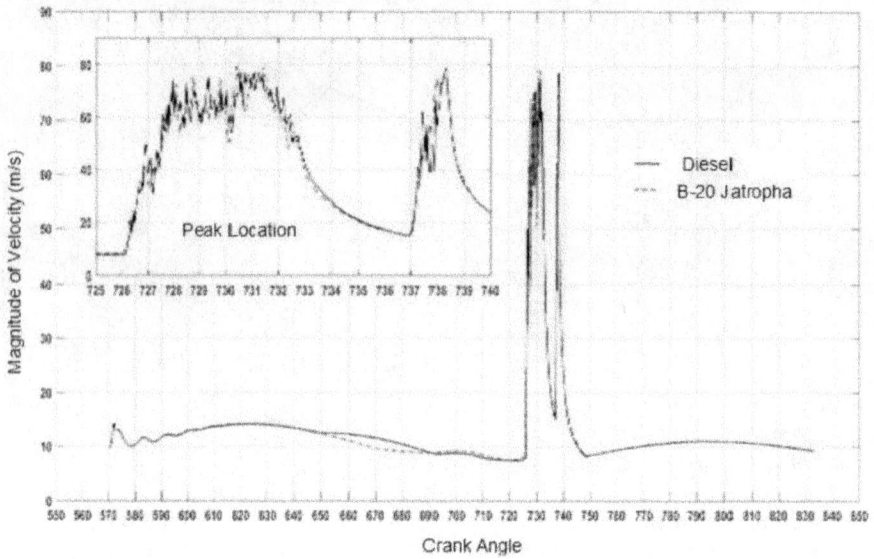

Figure 10.
Magnitude of velocity versus crank angle.

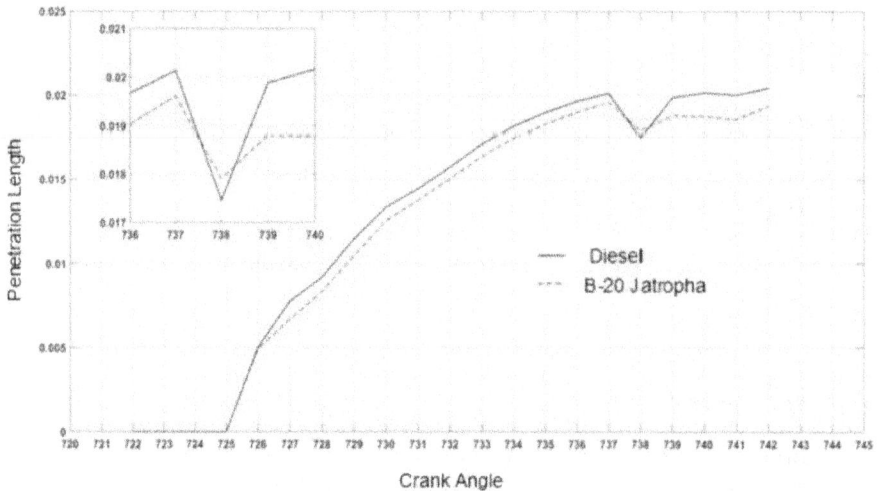

Figure 11.
Penetration length versus Crank angle.

Turbulence is of the major concern in the engine cylinder. The diffusion in the engine cylinder results from the local fluctuations in the flow field. This leads to the enhanced rates of momentum, heat and mass transfer yielding to the satisfactory engine operation. The engine flows involve complicated shear layer combination, boundary layer, and reticulating regions [27]. As the flow is unsteady it exhibits cycle by cycle fluctuations. In diesel engine swirl is used for rapid mixing between the inducted charge and the injected fuel. It is also used for speeding the combustion process. **Figure 12** shows the swirl ratio versus crank angle for diesel and B-20 Jatropha. For diesel and biodiesel compression starts at 570.25 deg. crank angle with a temperature of 404°C and ends at 712° CA with temperature 1008°C, Swirl ratio varies from 1.3 to 0.89 during the CA 570° to 830° for the biodiesel swirl ratio varies

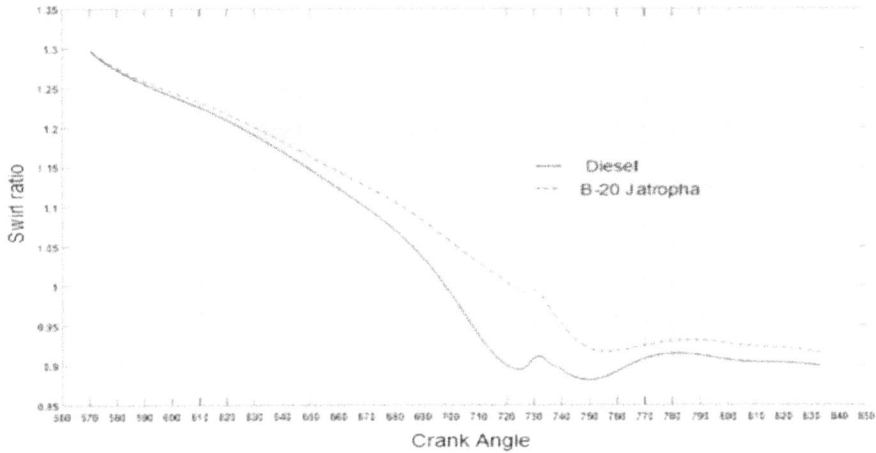

Figure 12.
Swirl ratio versus crank angle.

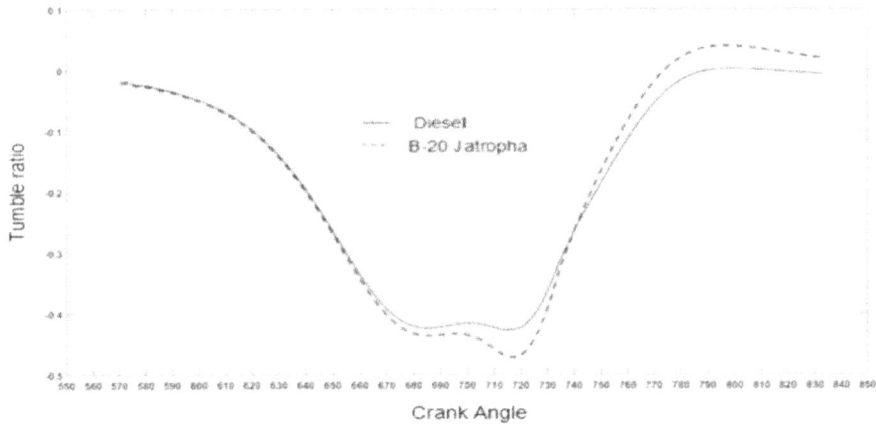

Figure 13.
Tumble ratio versus crank angle.

from 1.3 to 0.92 with same crank angle. B-20 JOME is having high viscosity when compared to that of diesel, which may lead to the complicated shear layer combination thereby increasing the thickness of the boundary layer.

Tumble ratio strongly affects the mixture formation. The high-pressure fuel injection certainly disturbs the bulk motion of the cylinder in the engine. The effect of fuel injection pressure on bulk motion of air is negligible because of symmetrical positioning of the fuel injector holes about the axis of the injector. The tumble ratio in **Figure 13** decreases (from −0.03 to −0.42) initially from 570° to 700° CA and again increases (from −0.42 to 0) during the combustion stroke (i.e., from 720 to 800° CA). In case of B-20 JOME, tumble ratio decreases from −0.03 to 0.44 and increases from 0.44 to 0.03. For biodiesel tumble ratio, there are some fluctuations from 700° to 720°CA. During that situation, piston is nearer to the TDC and fuel injection happens period. This phenomenon may be due to the presence of oxygen content in biodiesels leading to the oxidation process. This behaviour can be noticed from the temperature contour plot for diesel and biodiesel. Tumble ratio does not vary much till certain crank angle degree and for the reduced volume of high combustion chamber. Tumble ratio is found to be high at high engine speeds during the fuel injection phase because of high piston velocity helping tumble motion [28]. Peak pressure rise depends on the combustion rate during the initial phase. In turn it depends on the amount of fuel present in the uncontrolled combustion phase. The volatility of the slow-burning biodiesel increases the combustion duration thereby giving the high rate of pressure rise (see **Figure 14**).

Premixed burning phase associated with high heat release rate is significant to the diesel. It gives high thermal efficiency for the diesel. **Figure 15** shows apparent heat release rate (AHRR) versus crank angle. From the heat release rate graph, one can analyse the occurrence of short premixed heat release flame for the esters. Diffusion burning phase under the second peak is high for biodiesel when compared to that of diesel. This may be due to viscosity of biodiesels on fuel spray, reduction of air entrainment and fuel-air mixing rates. Biodiesels possess low latent heat of vaporization. Thereby, heat transfer lowers local air temperature [29]. The heat release rate for the biodiesel is found to be low when compared to that of diesel.

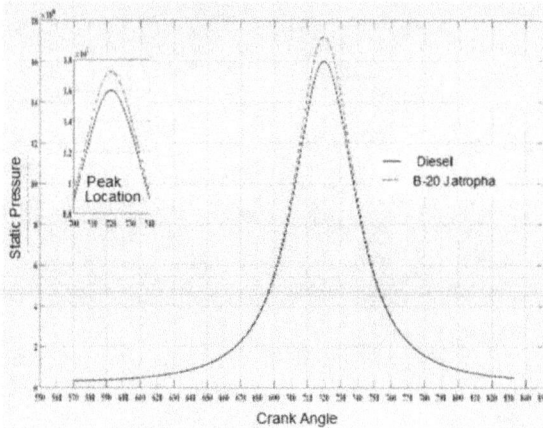

Figure 14.
Static pressure versus crank angle.

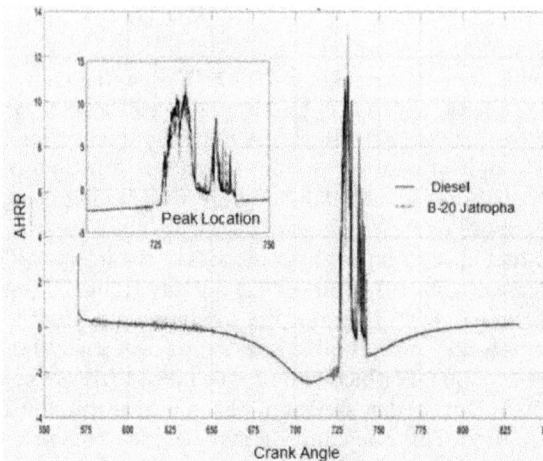

Figure 15.
Apparent heat release rate (AHRR) versus crank angle.

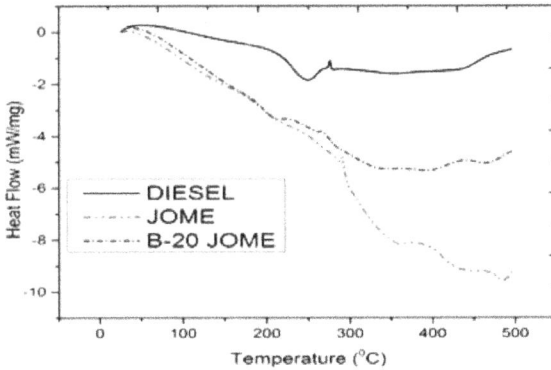

Figure 16.
Comparison of DSC Combustion curves.

Figure 16 shows the comparison of DSC combustion curves. Generally, combustion process of organic fuels exhibit exothermic reaction in air due to double bond presence [30]. JOME consists of the carbon number varying from 14 to 20 (i.e., C_{14} to C_{20}) which decomposes in the range of 30–240°C. JOME exhibits 298°C peak temperature of combustion with 84 J/g enthalpy. Biodiesel in engine results in hard burning with less enthalpy [31, 32]. Combustion curve of B-20 exhibits 268°C peak temperature with 147.5 J/g enthalpy, which is comparable to that of diesel having 138 J/g enthalpy. This indicates that combustion of B-20 JOME is close to that of diesel. Combustion of diesel molecules takes place initially followed by biodiesel [33, 34]. B-20 JOME combustion starts early resulting in better combustion when compared to JOME and diesel with high reaction region. During the initial phase of biodiesel combustion short pre-mixed flame occurs followed by diffusion burning phase requiring blending [34]. B-20 JOME indicates that JOME and diesel molecules mixed perfectly and homogenous mixture occurs at 20%. Therefore, performance of B-20 blend is close to that of diesel.

6. Conclusions

Combustion simulations are performed on the four-stroke single-cylinder direct injection compression ignition engine running at a constant speed of 1500 rpm, injection timing of 25° BTDC with diesel and 20% blend of Jatropha biodiesel. Standard FVM (finite volume method) of CFD (computational fluid dynamics) is considered while simulating the two-phase engine flow. 3-Dimensional N-S (Navier–Stokes) equations are solved with k-ε turbulence model.

- Results of combustion simulations are presented for half-cycle by considering the two strokes compression and expansion at zero load condition. *The physiochemical properties of B-20 blend differ significantly with those of diesel. The processes like fuel injection, mixture propagation, and combustion differ significantly. For* 570° to 830° CA, swirl ratio varies from 1.3 to 0.89 for diesel, whereas it varies from 1.3 to 0.92 for B-20 JOME. For 720°-800° CA, tumble ratio decreases from −0.03 to −0.42 and increases to zero during the combustion stroke. For B-20 JOME, tumble ratio decreases from −0.03 to −0.44 and increases to −0.03 for the same CA. Some fluctuations are observed in the biodiesel tumble ratio from 700° to 720°CA.

- Combustion simulations confirm the B-20 blend as an alternative for diesel. DSC profiles of the diesel and B-20 JOME show endothermic peak, which is related to vaporization of methyl esters for B-20 JOME and volatilization of small fraction for the diesel.

- Biodiesel exhibits high enthalpy despite combustion of the engine, and causing serious engine problems. B-20 blend exhibits high enthalpy when compared to that of diesel with reduced peak temperature. Biodiesel exhibits high enthalpy despite of satisfactory performance as a fuel, its high viscosity causing poor fuel atomization. The trend of simulations matches with DSC results.

Acknowledgements

The authors would like to acknowledge the encouragements received from the Visvesvarayya Technological University Belgaum and the Koneru Lakshmaiah Educational Foundation, deemed to be University, Vaddeswaram, India.

Conflict of interest

The authors declare that there is no conflict of interests.

Author details

Vinay Atgur[1,2], Gowda Manavendra[1], Gururaj Pandurangarao Desai[3] and Boggarapu Nageswara Rao[2*]

1 Department of Mechanical Engineering, Bapuji Institute of Engineering and Technology (BIET), Davangere, Karnataka, India

2 Department of Mechanical Engineering, Koneru Lakshmaiah Education Foundation (KLEF), Deemed to be University, Guntur, Andhra Pradesh, India

3 Department of Chemical Engineering, Bapuji Institute of Engineering and Technology (BIET), Davangere, Karnataka, India

*Address all correspondence to: bnrao52@rediffmail.com

IntechOpen

References

[1] Kumar N, Ram S. Performance and emission characteristics of biodiesel from different origins: A review. Renewable and Sustainable Energy Reviews. 2013;**21**:633-658. DOI: 10.1016/j.rser.2013.01.006

[2] Rajasekar E, Selvi S. Review of combustion characteristics of CI engines fueled with biodiesel. Renewable and Sustainable Energy Reviews. 2014;**35**: 390-399. DOI: 10.1016/j.rser.2014. 04.006

[3] Manaf ISA, Embong NH, Khazaai SNM, Rahim MHA, Yusoff MM, Lee KT, et al. A review for key challenges of the development of biodiesel industry. Energy Conversion and Management. 2019;**185**:508-517

[4] Jafarmadar S, Zehni A. Combustion modeling for modern direct injection diesel engines. Iranian journal of chemistry and chemical engineering. 2012;**31**(3):111-114

[5] Griend LV, Feldman ME, Peterson CL. Modeling combustion of alternative fuels in a DI diesel engine using KIVA. Transactions of the ASABE. 1988;**33**(2):342-350. DOI: 10.13031/ 2013.31336

[6] Cantrell BA, Reitz RD, Rutland CJ, Imamore Y. Strategies for reducing the computational time of diesel engine CFD simulations. In: International Multidimensional Engine Modeling User's Group Meeting. (SAE Congress, 23 April 2012), Warrendale, PA, USA: SAE International; 2012

[7] Fukuda K, Ghasemi A, Barron R, Balachandar R. An Open Cycle Simulation of DI Diesel Engine Flow Field Effect on Spray Processes. Warrendale, PA, USA: SAE International. SAE Technical Paper 2012-01-06962012. DOI: 10.4271/ 2012-01-0696

[8] Dimitriou P, Tsujimura T, Kojima H, Aoyagi K, Kurimoto N. Experimental and simulation analysis of natural gas-diesel combustion in dual-fuel engines. Frontiers in Mechanical Engineering. 2020;**6**:1-14. DOI: 10.3389/fmech.2020. 543808

[9] Baratta M, Chiriches S, Goel P, Misul D. CFD modelling of natural gas combustion in IC engines under different EGR dilution and H2-doping conditions. Transportation Engineering. 2020;2:1-12. DOI: 10.1016/j. treng.2020.100018

[10] Costa M, Piazzullo D. Biofuel powering of internal combustion engines: Production routes, effect on performance and CFD modeling of combustion. Frontiers of Mechanical Engineering. 2018;**4**:9. DOI: 10.3389/ fmech.2018.00009

[11] Maghbouli A, Khoshbakhti R, Shafee S, Ghafouri J. Numerical study of combustion and emission characteristics of dual-fuel engines using 3D-CFD models coupled with chemical kinetics. Fuel. 2013;**106**:98-105. DOI: 10.1016/j. fuel.2012.10.055

[12] Vijayashree GV. Application of CFD for analysis and design of IC engines. In: Srivastava D, Agarwal A, Datta A, Maurya R, editors. Advances in Internal Combustion Engine Research. Energy, Environment, and Sustainability. Singapore: Springer; 2018. DOI: 10.1007/978-981-10-7575-9_13

[13] Salam S, Choudhary T, Pugazhendhi A, Verma TN, Sharma A. A review on recent progress in computational and empirical studies of compression ignition internal combustion engine. Fuel. 2020;**279**: 118469. DOI: 10.1016/j.fuel.2020. 118469

[14] Kerstein AR. Turbulence in combustion processes: Modeling challenges. Proceedings of the Combustion Institute. 2002;**29**(2): 1763-1773. DOI: 10.1016/s1540-7489 (02)80214-0

[15] Zhou LX. Advances in studies on two-phase turbulence in dispersed multiphase flows. International Journal of Multiphase Flow. 2010;**36**(2): 100-108. DOI: 10.1016/j. ijmultiphaseflow.2009.02.011

[16] Azad AK, Rasul MG, Khan MMK, Sharma SC, Bhuiya MMK. Recent development of biodiesel combustion strategies and modelling for compression ignition engines. Renewable and Sustainable Energy Reviews. 2016;**56**:1068-1086. DOI: 10.1016/j.rser.2015.12.024

[17] Sasidhar Saketh AVS, Raja M, Nageswara RB. Selection of a suitable turbulent model in CFD codes to perform simulations for a circular microchannel heat exchanger. International Journal of Mechanical and Production Engineering Research and Development (IJMPERD), Special Issue. Aug 2018:608-623

[18] Basha SA, Gopal KR. In-cylinder fluid flow turbulence and spray models. Renewable and Sustainable Energy Reviews. 2008;**13**(6–7):1620-1627

[19] Sharma CS, Anand TNC, Ravikrishna RV. A methodology for analysis of diesel engine in-cylinder flow and combustion. Progress in Computational Fluid Dynamics. 2010; **10**(3):157-167. DOI: 10.1504/ PCFD.2010.033327

[20] Gugulothu SK, Reddy KHC. CFD simulation of in-cylinder flow on different piston bowl geometries in a DI diesel engine. Journal of Applied Fluid Mechanics. 2016;**9**(3):1147-1155. DOI: 10.18869/acadpub. jafm.68.228.24397

[21] Zellat M., Abouri D. and Conte T. (2014). Advanced Modeling of DI Diesel Engines Investigations on Combustion, High EGR Level and Multiple- Injection Application to DI Diesel Combustion Optimization. https://www.energy.gov/ sites/prod/files/2014/03/f9/2005_deer_ zellat.pdf

[22] Posom J, Sirisomboon P. Evaluation of the thermal properties of Jatropha curcas L. kernels using near-infrared spectroscopy. Biosystems Engineering. 2014;**125**:45-53. DOI: 10.1016/j. biosystemseng.2014.06.011

[23] Kongre UV, Sunnapwar VK. CFD modeling and experimental validation of combustion in direct ignition engine fueled with diesel. International Journal of Applied Engineering Research. 2010; **1**(3):508-517

[24] Imamore Y, Hiraoka K. Combustion simulations contributing to the development of reliable low-emission diesel engines. Mitsubishi Heavy. 2011; **48**(1):65-69 http://www.mhinglobal.c om/company/technology/review/pdf/e 481/e481065.pdf

[25] Paul G, Datta A, Mandal BK. An experimental and numerical investigation of the performance, combustion and emission characteristics of a diesel engine fueled with jatropha biodiesel. Energy Procedia. 2014;**54**: 455-467. DOI: 10.1016/j. egypro.2014.07.288

[26] Dembinski HWR. In-cylinder Flow Characterisation of Heavy Duty Diesel Engines Using Combustion Image Velocimetry. Mumbai, India: Integrated Publishing Association. 2013. DOI: 10.1016/j.radonc. 2014.09.010

[27] Mirmohammadi A, Ommi F. Internal combustion engines in-cylinder flow simulation improvement using nonlinear k-ε turbulence models. Journal of Computational and Applied

Research in Mechanical Engineering. 2015;**5**(1):61-69

[28] Raj RTK, Manimaran R. Effect of swirl in a constant speed DI diesel. CFD Letters. 2012;**4**:214-224

[29] Banapurmath NR, Tewari PG. Performance, combustion, and emissions characteristics of a single-cylinder compression ignition engine operated on ethanol-biodiesel blended fuels. Proceedings of the Institution of Mechanical Engineers, Part A: Journal of Power and Energy. 2010;**224**(4): 533-543. DOI: 10.1243/09576509JPE850

[30] Mohammed MN, Atabani AE, Uguz G, Lay CH, Kumar G, Al-Samaraae RR. Characterization of Hemp (Cannabis sativa L.) biodiesel blends with Euro diesel, butanol and diethyl ether using FT-IR, UV–Vis, TGA and DSC techniques. Waste and Biomass Valorization. 2020;**11**(3):1097–1113. DOI: 10.1007/s12649-018-0340-8

[31] Dinkov R, Hristov G, Stratiev D, Boynova AV. Effect of commercially available antioxidants over biodiesel/diesel blends stability. Fuel. 2009;**88**(4): 732-737. DOI: 10.1016/j. fuel.2008.09.017

[32] Vossoughi S, El-Shoubary YM. Kinetics of liquid hydrocarbon combustion using the DSC technique. Thermochimica Acta. 1990;**157**(1): 37-44. DOI: 10.1016/0040-6031(90) 80004-I

[33] Vinay Atgur G, Manavendra GPD, Nageswara Rao B. Thermal characterization of dairy washed scum methyl ester and its b-20 blend for combustion applications. International Journal of Ambient Energy. 2021 (in Press). DOI: 10.1080/ 01430750.2021.1909651

[34] Almazrouei M, Janajreh I. Thermogravimetric study of the combustion characteristics of biodiesel

and petroleum diesel. Journal of Thermal Analysis and Calorimetry. 2019;**136**(2):925-935. DOI: 10.1007/ s10973-018-7717-6

Comparison of CFD and FSI Simulations of Blood Flow in Stenotic Coronary Arteries

Violeta Carvalho, Diogo Lopes, João Silva, Hélder Puga,
Rui A. Lima, José Carlos Teixeira and Senhorinha Teixeira

Abstract

Cardiovascular diseases are amongst the main causes of death worldwide, and the main underlying pathological process is atherosclerosis. Over the years, fatty materials are accumulated in the arterial which consequently hinders the blood flow. Due to the great mortality rate of this disease, hemodynamic studies within stenotic arteries have been of great clinical interest, and computational methods have played an important role. Commonly, computational fluid dynamics methods, where only the blood flow behavior is considered, however, the study of both blood and artery walls' interaction is of foremost importance. In this regard, in the present study, both computational fluid dynamics and fluid-structure interaction modeling analysis were performed in order to evaluate if the arterial wall compliance affects considerably the hemodynamic results obtained in idealized stenotic coronary models. From the overall results, it was observed that the influence of wall compliance was noteworthy on wall shear stress distribution, but its effect on the time-averaged wall shear stress and on the oscillatory shear index was minor.

Keywords: atherosclerosis, blood flow, coronary arteries, fluid-structure interaction, computational fluid dynamics

1. Introduction

Over the last year, COVID-19 has been the most widely spoken and researched disease worldwide and, inevitably, other existing pathologies were moved to the background. Amongst these, cardiovascular diseases (CVDs) should be highlighted because they still are by far the major contributor to global mortality [1]. CVDs are mainly caused by a blockage that prevents blood from flowing properly, known as atherosclerosis [2, 3]. This is a complex disease that affects medium and large size arteries and consists of a build-up of fatty deposits on the inner walls of the blood vessels, hampering the blood flow through the body [4]. The effect of atherosclerosis can be exacerbated by other diseases that affect blood circulation. Particularly, it has been shown that patients with COVID-19 are prone to develop blood clots on both arteries and veins [5], and thus atherosclerosis may be an even more important factor to global mortality.

Given the prevalence of this disease, atherosclerosis has been intensely studied through both cardiovascular modeling and experimental procedures, as reviewed elsewhere [6–9]. Nevertheless, with the growing trend of greater computer power, computational approaches have become a valuable, cheaper, and efficient alternative for numerous researchers to predict blood flow behavior [10–13].

There are two main approaches for simulating blood flow, computational fluid dynamics (CFD) and fluid-structure interaction (FSI). In the first one, the arterial wall is assumed as rigid, while in the second one, arteries are considered elastic and the interaction between the blood and the arterial walls are included in the simulation [14]. Although CFD has been widely applied in the study of blood flow under pathological conditions in virtue of lower computational cost [15–20], since FSI provides a more realistic simulation of the human vasculature behavior, it has received increasing interest [21, 22]. Nonetheless, this approach requires significantly more computational effort, and the foremost difficulty is stability and convergence [23, 24].

In order to evaluate if the differences between CFD and FSI results are significant, some researchers have investigated and compared both. A commonly mentioned work was developed by Torii et al. [25]. The authors studied the effects of wall compliance on a stenotic patient-specific coronary artery and found noticeable differences in the instantaneous wall shear stress (WSS) produced by the FSI and rigid wall models. However, the effects of wall compliance on time-averaged WSS (TAWSS) and oscillatory shear index (OSI) were negligible. Malvè et al. [26] performed a similar study in the left coronary artery bifurcation and the conclusions regarding the WSS agreed with the previous study, but they observed significant differences in the TAWSS, especially on its spatial distribution. More recently, another similar study was performed on carotid bifurcation [10]. The authors found that the rigid model overestimates the flow velocity and WSS, but its influence on the TAWSS is minimal.

The differences between the CFD and FSI simulations have thus been the subject of several studies, however, as demonstrated, there is still a debate as to whether it is really necessary to use the most realistic approach, namely in computing WSS dependent variables, and sometimes the findings are contradictory. In this regard, this work presents the comparison of the results of both CFD and FSI simulations in an idealized stenotic coronary artery, with a degree of stenosis of 50%.

2. Methods

The first step necessary to study blood flow consisted in obtaining the three-dimensional geometry of the lumen of an idealized coronary artery as depicted in **Figure 1**. This geometry was previously adopted in other works by these authors [27, 28].

In the construction of the solid domain, a thickness of 0.8 mm was considered, according to an *in vivo* study [29].

Moreover, the fluid domain and the solid domain were discretized in 428,800 and 15,480 hexahedral elements, respectively, making a total of 444,280 elements for the FSI simulations. To ensure the quality of the mesh, the skewness and orthogonal quality were evaluated. The average skewness and orthogonal quality were approximately $8.3\,e^{-2}$ and 0.98, respectively. These parameters prove that the mesh is reliable for the study presented [30] and it is worth mentioning that a previous mesh study was carried out for CFD simulations [16], which was then adapted for this study.

2.1 Mathematical formulation

Blood flow is governed by the incompressible Navier-Stokes and the continuity equations as described in Eqs. (1) and (2),

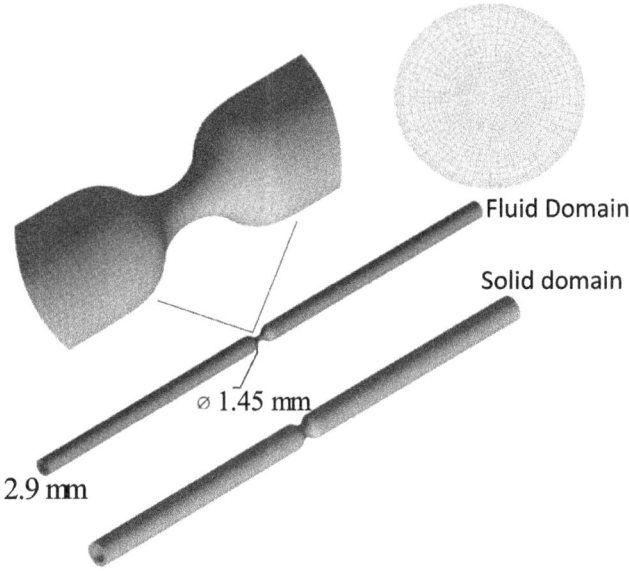

Figure 1.
Geometry and mesh of the coronary artery for both solid and fluid domain with 50% of stenosis.

$$\nabla \cdot u = 0 \tag{1}$$

$$\rho\left(\frac{\partial u}{\partial t} + u \cdot \nabla\right) = -\nabla p + \mu \nabla^2 u \tag{2}$$

where u is the velocity, p is the static pressure, ρ is the fluid density, and μ the dynamic viscosity [30, 31].

In the present study, blood was modeled as incompressible, laminar, and non-Newtonian fluid, having a density of 1060 kg/m³ [32]. Although it is commonly assumed as a Newtonian fluid, the ability of red blood cells to deform and aggregate makes blood a non-Newtonian fluid [33]. The well-known Carreau model was used to simulate the shear-thinning blood behavior, and it is defined by Eq. (3) [30, 31]:

$$\mu = \mu_\infty + (\mu_0 - \mu_\infty)\left[1 + \lambda\dot{\gamma}^2\right]^{\frac{n-1}{2}} \tag{3}$$

where μ is the viscosity, $\mu_\infty = 0.00345$ Pa•s is the infinite shear viscosity, $\mu_0 = 0.0560$ Pa · s is the blood viscosity at zero shear rate, $\dot{\gamma}$ is the instantaneous shear rate, $\lambda = 3.313$ s is the time constant and $n = 0.3568$ is the power-law index, as previously applied by [27].

The governing equation for the solid domain is the equilibrium equation (Eq. (4)) [30, 31]:

$$\rho_s\frac{\partial^2 u}{\partial t^2} - \nabla \bar{\bar{\sigma}} = \rho_s\vec{b} \tag{4}$$

where ρ_s is the solid density, u represents the solid displacements, $\vec{b}$, the body forces applied on the structure, and $\bar{\bar{\sigma}}$ is the Cauchy stress tensor. For an isotropic linear elastic solid, the stress tensor is represented by Eq. (5) [30, 31]:

$$\bar{\bar{\sigma}} = 2\,\mu_L\,\bar{\bar{\varepsilon}} + \lambda_L tr\left(\bar{\bar{\varepsilon}}\right)I \tag{5}$$

where λ_L and μ_L are the first and second Lamé parameters, respectively, $\bar{\bar{\varepsilon}}$, the strain tensor, tr, the trace function, and I, the identity matrix. For compressible materials, Lamé parameters can be written as a function of Young's modulus, E, and Poisson's coefficient, v, as follows.

$$\lambda_L = \frac{vE}{(1+v)(2v-1)} \tag{6}$$

$$\mu_L = \frac{E}{2\,(1+v)} \tag{7}$$

The arterial wall was modeled as a linear elastic, incompressible, isotropic, and homogeneous material with Young's modulus of 3.77 MPa [34], a density of 1120 kg/m³ [35], and a Poisson's ratio of 0.49 [21].

The FSI simulations were performed using the Arbitrary-Lagrangian-Eulerian (ALE) methodology for the fluid flow. Taking into account that the interface between the lumen and the wall deforms, the equations governing fluid flow have to be expressed in terms of the fluid variables relative to the mesh movement. The ALE-modified Navier-Stokes momentum equation for a viscous incompressible flow is described as follows in Eq. (8) [30, 31]:

$$\rho_f\left(\frac{\partial u}{\partial t} + \left((u - u_g) \cdot \nabla\right)u\right) = -\nabla p + \mu\nabla^2 u \tag{8}$$

where $\rho_f, p, u,$ and u_g are the fluid density, the pressure, the fluid velocity, and the moving coordinate velocity, respectively. The term $(u - u_g)$, in the ALE formulation, is added to the conventional Navier-Stokes equations to account for the movement of the mesh.

The displacement and equilibrium forces at the interface are represented by Eqs. (9) and (10) [30, 31]:

$$u_{f,\Gamma} = u_{s,\Gamma} \tag{9}$$

$$\overrightarrow{t_{f,\Gamma}} = \overrightarrow{t_{s,\Gamma}} \tag{10}$$

where $u_{f,\Gamma}$ is the displacement of the fluid at the interface, $u_{s,\Gamma}$, the displacement of the solid at the interface, $\overrightarrow{t_{f,\Gamma}}$, the forces of the fluid on the interface and $\overrightarrow{t_{s,\Gamma}}$, the forces of the solid on the interface.

2.2 Boundary conditions

Regarding the boundary conditions used in this study, at the inlet, a physiologically accurate pulsatile velocity profile was set, which is depicted in **Figure 2**. At the outlet, a pressure of 80 mmHg was assumed [17, 36].

The solid and fluid wall-boundaries were defined as a fluid-structure interface, and the inlet/outlet adjacent solid boundaries were fixed in all directions.

2.3 Numerical solution

For the CFD simulations, the Ansys Fluent software was used which applies the finite-volume discretization method. In this method, the fluid domain is divided

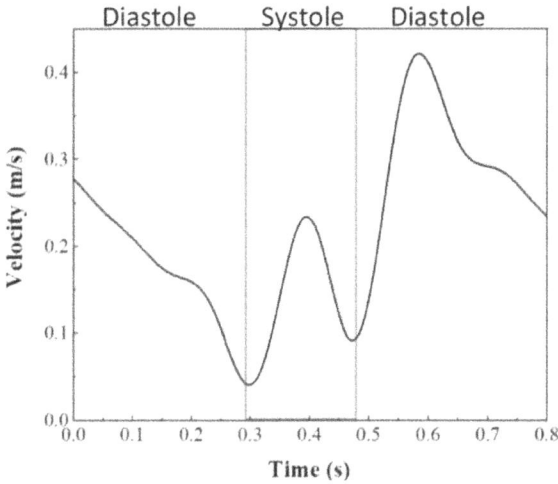

Figure 2.
Velocity profile implemented in CFD and FSI simulations.

into a finite number of control volumes, the conservation equations are applied to each control volume. Then, a system of algebraic equations for the variables is obtained. For the velocity-pressure coupling, the semi-implicit method for the pressure-linked equations (SIMPLE) scheme was used [37].

In FSI simulations, the same finite-volume method is applied in the fluid domain, and the computed forces in Fluent are transferred to the solid domain, through the interface. The finite element method (FEM) is used to solve the governing equations of the solid domain. Then, the computed displacements are transferred back to the fluid domain. This two-way coupling process was repeated until the difference of the displacements and forces for the last two iterations is below 1%. A time step of 0.01 s was used for every simulation.

2.4 Hemodynamic parameters

The formation of atherosclerotic lesions has been widely associated with hemo-dynamic parameters, such as the wall shear stress (WSS) and its indices, time-averaged wall shear stress (TAWSS), and oscillatory shear index (OSI) [38, 39]. These have been very useful to predict and estimate disturbed flow conditions and the development of local atherosclerotic plaques [27, 40].

The spatial WSS, τ_w, is calculated by Eq. (11), being $\dot{\gamma}$, the deformation rate, and μ the dynamic viscosity.

$$\tau_w = \mu \frac{\partial u}{\partial y} = \mu \, \dot{\gamma} \tag{11}$$

The TAWSS index allows obtaining an average temporal evaluation of the WSS exerted during a cardiac cycle (T) [40]. This is calculated by Eq. (12):

$$TAWSS = \frac{1}{T} \int_0^T |WSS| dt \tag{12}$$

The OSI index is the temporal fluctuation of low and high average shear stress during a cardiac cycle (T) and it is calculated by applying Eq. (13):

$$OSI = \frac{1}{2}\left(1 - \frac{\left|\int_0^T WSSdt\right|}{\int_0^T |WSS|dt}\right)$$

(13)

The formulation developed in this section describes the models that couple the fluid dynamics and the mechanical interaction with the arteries' wall which is treated as a deformable material. This methodology enables the computation of critical parameters for understanding the hemodynamics in the presence of a stenosis, such as the WSS. The advantages of this method are made evident by comparing it with a simple CFD analysis as detailed in the following section.

3. Results and discussion

3.1 Velocity profiles along the cardiac cycle

Figure 3 illustrates the velocity profiles measured along the center of the artery for both CFD and FSI simulations, in two different phases of the cardiac cycle: systole (0.4 s) and diastole (0.58 s).

Looking at the results in **Figure 3**, it can be observed that the estimated velocity profiles are similar for both CFD and FSI modeling. Moreover, as expected, during diastole (**Figure 3b**) the velocities measured are higher than during systole (**Figure 3a**). This happens because, during systole, the coronary arteries are compressed by the contraction of the myocardium, and so, most of the coronary flow occurs during diastole, where the flow increases. In addition, the maximum velocities in both cases are measured in the stenosis throat ($x = 0$ mm) as observed by other investigations [15, 27].

It is also noted that the velocity is overestimated in the CFD model, particularly at and upstream of the stenosis throat. This is expected because the deformation of the elastic model provides a larger volume for the blood flow through and, as the inflow rate is equal for both models, naturally, the velocities will be higher when a rigid wall is considered. Downstream of the stenosis, at an x coordinate of approximately 10 mm, the velocity is higher for the FSI case, which indicates that the pressure drop at the stenosis creates zones of low pressure, which contracts the artery, and forces the flow to accelerate. In this case, it is thus observed the effect of artery compliance, which allows a steadier supply of flow despite the variable nature of the cardiac cycle.

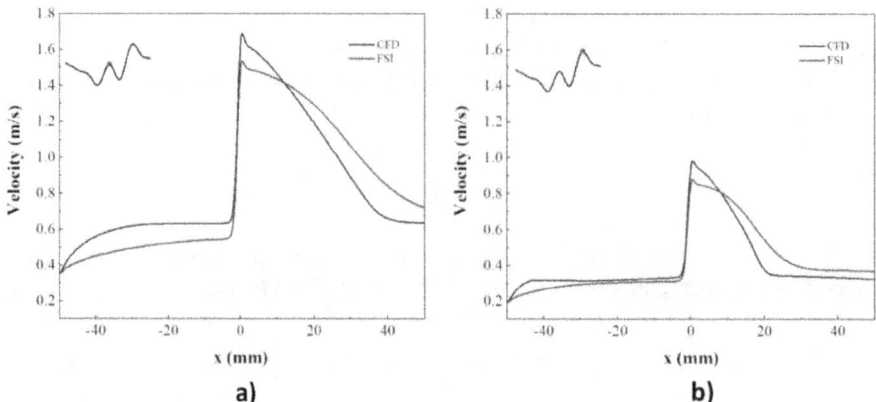

Figure 3.
Axial flow velocity profiles at the center line drawn across the artery at (a) systole and (b) diastole.

3.2 Velocity streamlines along the cardiac cycle

After evaluating the velocity profiles, velocity streamlines were created to better visualize and understand how blood flow behaves, as shown in **Figure 4**.

The results indicate the existence of fluid recirculation downstream of the stenosis for both CFD and FSI models, but the recirculation zones are longer in FSI simulations, and this is due to the considerable vessel expansion driven by the pulsatile blood flow.

3.3 WSS and its indices

The magnitude of the WSS predicted by FSI and rigid model along the artery wall for both systole and diastole are compared in **Figure 5**. In this case, it is observed that the differences between compliant and rigid-wall models are remarkable. In CFD simulations, WSS values estimated in the stenosis throat are approximately twice of those obtained with FSI simulations. This indicates that the WSS distributions were substantially affected by arterial wall compliance, which is in agreement with previous research [25, 26, 41].

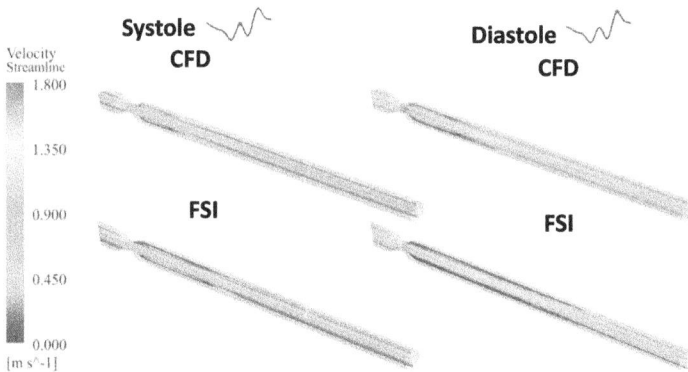

Figure 4.
Velocity streamlines obtained during systole and diastole for both CFD and FSI simulations.

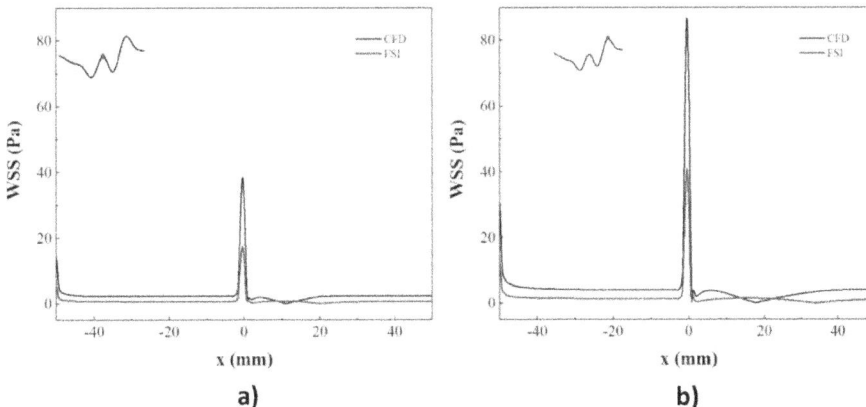

Figure 5.
Wall shear stress profiles obtained along artery wall at (a) systole and (b) diastole.

Taking into account that WSS-related hemodynamic parameters, such as OSI and TAWSS, play an important role in atherogenesis, these were also evaluated. **Figure 6** depicts the TWASS profiles obtained in both CFD and FSI simulations.

The results evidence high values of TAWSS at the stenosis throat, due to flow acceleration and high-velocity gradient near the wall, and, as expected the TAWSS is overestimated by the CFD model as previously explained for WSS distributions. In spite of these observations, the overall TAWSS distributions for the FSI and rigid-wall cases are identical.

Regarding the OSI profiles depicted in **Figure 7**, it can be observed that for both CFD and FSI simulations, the maximum values (≈ 0.5) are obtained downstream of the stenosis, which indicates the presence of unsteady and oscillatory flow, commonly associated with higher susceptibility to atherosclerotic plaque development [27, 40]. Nonetheless, although the OSI profiles for the two cases look similar and unaffected by wall distensibility, in the distal region, OSI values for the FSI case are slightly higher than for the rigid wall [25]. These differences may be due to the occurrence of longer recirculation areas with the elastic model.

In general, it was found that for both rigid and compliant models the post-stenotic region presents lower TAWSS and higher OSI values, which constitute risk factors for the incidence and abnormal plaque formation [16, 27].

3.4 Displacement

Figure 8 represents the arterial wall displacement contours of the FSI simulations during systole and diastole.

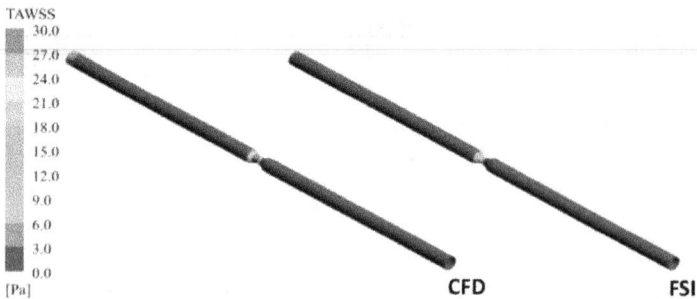

Figure 6.
Time-averaged wall shear stress profiles obtained in CFD and FSI simulations.

Figure 7.
Oscillatory shear index obtained in CFD and FSI simulations.

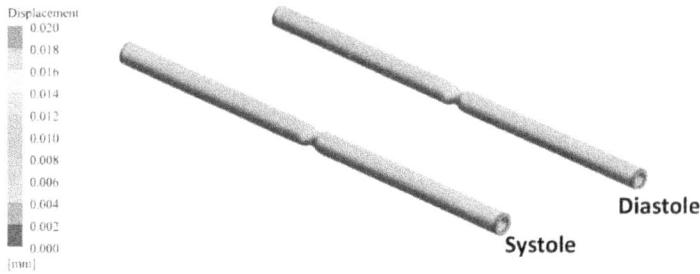

Figure 8.
Displacement profile at the arterial wall obtained in CFD and FSI simulations.

In the first place, it is noted that the displacements are similar for both cycle phases. This is a consequence of the assumption of the constant outlet pressure, which is a limitation of this work. In this case, the deformations are slightly bigger in the region upstream of the stenosis, and as there is a pressure drop in the throat, the displacements are somewhat lower in the downstream region.

It is also noteworthy that the displacements of the arterial wall are approximately 1.5–2% of the vessel thickness, which is in agreement with the hypothesis that arteries become stiffer with the development of atherosclerosis. Despite this order of magnitude of the displacement values, these still bring considerable differences in the calculations of the velocity and WSS-related variables.

4. Conclusions

The study of blood flow in stenotic arteries has been made mainly assuming the artery's wall as rigid, however, in reality, they are naturally elastic. Given that there are some inconsistencies in the literature regarding the comparison between CFD and FSI simulations, in the present work, a comparative study between FSI and CFD modeling was performed in order to investigate the influence of artery compliance on stenotic coronary artery hemodynamics and wall shear stress distribution. The main conclusions obtained from this work are:

- Comparing the rigid and compliant models, the velocities predicted differ slightly.

- The difference between WSS profiles was remarkable. The CFD simulations overestimate the WSS values, which consequently indicates that when a more realistic WSS estimation is needed, is essential to consider the effect of the atrial wall on blood flow.

- Insignificant differences were verified in the TAWSS and OSI measurements.

Acknowledgements

This work has been supported by FCT—Fundação para a Ciência Tecnologia within the R&D Units Project Scope: UIDB/00319/2020, UIDB/04077/2020, UIDB/04436/2020, and NORTE-01-0145-FEDER-030171 (PTDC/EMD-EMD/30171/2017) and EXPL/EME-EME/0732/2021, NORTE-01-0145-FEDER-029394 funded by COMPETE2020, NORTE 2020, PORTUGAL 2020. Violeta Carvalho,

Diogo Lopes and João Silva would like to express their gratitude for the support given by FCT through the PhD Grants SFRH/UI/BD/151028/2021, SFRH/BD/144431/2019, and SFRH/BD/130588/2017, respectively.

Conflict of interest

The authors declare no conflict of interest.

Author details

Violeta Carvalho[1,2*], Diogo Lopes[3], João Silva[1,2], Hélder Puga[3], Rui A. Lima[1,4], José Carlos Teixeira[1] and Senhorinha Teixeira[2]

1 MEtRICs, University of Minho, Guimarães, Portugal

2 ALGORITMI, University of Minho, Guimarães, Portugal

3 Center for MicroElectromechanical Systems (CMEMS-UMinho), University of Minho, Guimarães, Portugal

4 CEFT, Faculty of Engineering of the University of Porto, Porto, Portugal

*Address all correspondence to: violeta.carvalho@dem.uminho.pt

IntechOpen

References

[1] Mendis S, Puska P, Norrving B. Global Atlas of Cardiovascular Disease Prevention and Control. Geneva, Switzerland: World Health Organization; 2011. ISBN: 9789241564373

[2] World Health Organization (WHO). Cardiovascular Diseases (CVDs): Fact Sheet. 2017. Available from: https://www.who.int/news-room/fact-sheets/detail/cardiovascular-diseases-(cvds)

[3] Kim J, Jin D, Choi H, Kweon J, Yang DH, Kim YH. A zero-dimensional predictive model for the pressure drop in the stenotic coronary artery based on its geometric characteristics. Journal of Biomechanics. 2020;113:110076. DOI: 10.1016/j.jbiomech.2020.110076

[4] Shah PK. Inflammation, infection and atherosclerosis. Trends in Cardiovascular Medicine. 2019;29:468-472. DOI: 10.1016/j.tcm.2019.01.004

[5] Zuo Y et al. Prothrombotic autoantibodies in serum from patients hospitalized with COVID-19. Science Translational Medicine. 2020;3876:1-17. DOI: 10.1126/scitranslmed.abd3876

[6] Carvalho V, Pinho D, Lima RA, Teixeira JC, Teixeira S. Blood flow modeling in coronary arteries: A review. Fluids. 2021;6(2):53. DOI: 10.3390/fluids6020053

[7] Lopes D, Puga H, Teixeira J, Lima R. Blood flow simulations in patient-specific geometries of the carotid artery: A systematic review. Journal of Biomechanics. 2020;111:110019. DOI: 10.1016/j.jbiomech.2020.110019

[8] Carvalho V et al. In vitro stenotic arteries to perform blood analogues flow visualizations and measurements: A review. Open Biomedical Engineering Journal. 2020;14:87-102. DOI: 10.2174/1874120702014010087

[9] Zhang J-M et al. Perspective on CFD studies of coronary artery disease lesions and hemodynamics: A review. International Journal of Numerical Methods in Biomedical Engineering. 2014;30(6):659-680. DOI: 10.1002/cnm

[10] Lopes D, Puga H, Teixeira JC, Teixeira SF. Influence of arterial mechanical properties on carotid blood flow: Comparison of CFD and FSI studies. International Journal of Mechanical Sciences. 2019;160:209-218. DOI: 10.1016/j.ijmecsci.2019.06.029

[11] Karimi A, Navidbakhsh M, Razaghi R, Haghpanahi M. A computational fluid-structure interaction model for plaque vulnerability assessment in atherosclerotic human coronary arteries. Journal of Applied Physics. 2014;115(14):144702-8. DOI: 10.1063/1.4870945

[12] Rammos KS, Koullias GJ, Pappou TJ, Bakas AJ, Panagopoulos PG, Tsangaris SG. A computer model for the prediction of left epicardial coronary blood flow in normal, stenotic and bypassed coronary arteries, by single or sequential grafting. Cardiovascular Surgery. 1998;6(6):635-648. DOI: 10.1177/096721099800600617

[13] Pincombe B, Mazumdar J. The effects of post-stenotic dilatations on the flow of a blood analogue through stenosed coronary arteries. Mathematical and Computer Modelling. 1997;25(6):57-70. DOI: 10.1016/S0895-7177(97)00039-3

[14] Siogkas PK et al. Patient-specific simulation of coronary artery pressure measurements: An in vivo three-dimensional validation study in humans. BioMed Research International. 2014;2015:628416. DOI: 10.1155/2015/628416

[15] Carvalho V, Rodrigues N, Ribeiro R, Costa PF, Lima RA, Teixeira SFCF. 3D printed biomodels for flow visualization in stenotic vessels: An experimental and numerical study. Micromachines. 2020; **11**(6):549. DOI: 10.3390/mi11060549

[16] Carvalho V et al. Hemodynamic study in 3D printed stenotic coronary artery models: Experimental validation and transient simulation. Computer Methods in Biomechanics and Biomedical Engineering. 2020;**24**:623-636. DOI: 10.1080/10255842.2020.1842377

[17] Kamangar S et al. Effect of stenosis on hemodynamics in left coronary artery based on patient-specific CT scan. Bio-medical Materials and Engineering. 2019;**30**(4):463-473. DOI: 10.3233/BME-191067

[18] Liu H et al. Effect of microcirculatory resistance on coronary blood flow and instantaneous wave-free ratio: A computational study. Computer Methods and Programs in Biomedicine. 2020;**196**: 105632. DOI: 10.1016/j.cmpb.2020.105632

[19] Narayan S, Saha S. Time-dependent study of blood flow in an aneurysmic stenosed coronary artery with inelastic walls. Materials Today: Proceedings. 2021;**47**:4718–4724. DOI: 10.1016/j. matpr.2021.05.608

[20] Zhao Y, Ping J, Yu X, Wu R, Sun C, Zhang M. Fractional flow reserve-based 4D hemodynamic simulation of time-resolved blood flow in left anterior descending coronary artery. Clinical biomechanics. 2019;**70**:164-169. DOI: 10.1016/j.clinbiomech.2019.09.003

[21] Bukač M, Čanić S, Tambača J, Wang Y. Fluid–structure interaction between pulsatile blood flow and a curved stented coronary artery on a beating heart: A four stent computational study. Computer Methods in Applied Mechanics and Engineering. 2019;**350**:679-700. DOI: 10.1016/j.cma.2019.03.034

[22] Arefin MS. Hemodynamic and structural effects on bypass graft for different levels of stenosis using fluid structure interaction: A prospective analysis. Journal of Vascular Nursing. 2019;**37**(3):169-187. DOI: 10.1016/j. jvn.2019.05.006

[23] Ahmadi M, Ansari R. Computational simulation of an artery narrowed by plaque using 3D FSI method: Influence of the plaque angle, non-Newtonian properties of the blood flow and the hyperelastic artery models. Biomedical Physics & Engineering Express. 2019;**5**(4):45037. DOI: 10.1088/2057-1976/ab323f

[24] Zhang X, Luo M, Wang E, Zheng L, Shu C. Numerical simulation of magnetic nano drug targeting to atherosclerosis: Effect of plaque morphology (stenosis degree and shoulder length). Computer Methods and Programs in Biomedicine. 2020;**195**: 105556. DOI: 10.1016/j.cmpb.2020. 105556

[25] Torii R et al. Fluid–structure interaction analysis of a patient-specific right coronary artery with physiological velocity and pressure waveforms. Communications in Numerical Methods in Engineering. 2009;**25**(25):565-580. DOI: 10.1002/cnm

[26] Malvè M, García A, Ohayon J, Martínez MA. Unsteady blood flow and mass transfer of a human left coronary artery bifurcation: FSI vs. CFD. International Communications in Heat and Mass Transfer. 2012;**39**(6):745-751. DOI: 10.1016/j.icheatmasstransfer. 2012.04.009

[27] Carvalho V, Rodrigues N, Lima RA, Teixeira SFCF. Modeling blood pulsatile turbulent flow in stenotic coronary arteries. International Journal of Biology and Biomedical Engineering. 2020;**14**(22): 1998-4510. DOI: https://doi.org/ 10.46300/91011.2020.14.22

[28] Carvalho V, Rodrigues N, Lima RA, Teixeira S. Numerical simulation of blood pulsatile flow in stenotic coronary arteries: The effect of turbulence modeling and non-Newtonian assumptions. In: International Conference on Applied Mathematics & Computer Science. 2020. pp. 112-116. DOI: 10.1109/CSCC49995.2020.00027

[29] Fayad ZA et al. Noninvasive in vivo human coronary artery lumen and wall imaging using black-blood magnetic resonance imaging. Circulation. 2000;**102**:506-510

[30] Ansys I. ANSYS® Fluent User's Guide, Release 2020 R2. Canonsburg: ANSYS, Inc. 2020

[31] Ansys I. ANSYS® Fluent Theory Guide, Release 2020 R2. Canonsburg: ANSYS, Inc; 2020

[32] Wu X, von Birgelen C, Zhang S, Ding D, Huang J, Tu S. Simultaneous evaluation of plaque stability and ischemic potential of coronary lesions in a fluid–structure interaction analysis. The International Journal of Cardiovascular Imaging. 2019;**35**(9): 1563-1572. DOI: 10.1007/s10554-019-01611-y

[33] Boujena S, El Khatib N, Kafi O. Generalized Navier–stokes equations with non-standard conditions for blood flow in atherosclerotic artery. Applicable Analysis. 2016;**95**(8): 1645-1670. DOI: 10.1080/00036811.2015.1068297

[34] Karimi A, Navidbakhsh M, Shojaei A, Faghihi S. Measurement of the uniaxial mechanical properties of healthy and atherosclerotic human coronary arteries. Materials Science and Engineering: C. 2013;**33**(5):2550-2554. DOI: 10.1016/j.msec.2013.02.016

[35] Sousa LC, Castro CF, António CC, Azevedo E. Fluid-Structure Interaction Modeling of Blood Flow in a

Non-Stenosed Common Carotid Artery Bifurcation. In: Proceedings of the Proceedings of the 7th International Conference on Mechanics and Materials in Design; Silva Gomes JF, Meguid S, Eds.; INEGI/FEUP, 2017; pp. 1559–1564

[36] Rabbi MF, Laboni FS, Arafat MT. Computational analysis of the coronary artery hemodynamics with different anatomical variations. Informatics in Medicine Unlocked. 2020;**19**:100314. DOI: 10.1016/j.imu.2020.100314

[37] Ferziger JH, Peric M. Computational Methods for Fluid Dynamics, Third. New York City: Springer; 2002

[38] Han D et al. Relationship between endothelial wall shear stress and high-risk atherosclerotic plaque characteristics for identification of coronary lesions that cause ischemia: A direct comparison with fractional flow reserve. Journal of the American Heart Association. 2016;**5**(12):1-10. DOI: 10.1161/JAHA.116.004186

[39] Ku D. Blood flow in arteries. Annual Review of Fluid Mechanics. 1997;**29**: 399-434. DOI: 10.1146/annurev.fluid.29.1.399

[40] Pinto SIS, Campos JBLM. Numerical study of wall shear stress-based descriptors in the human left coronary artery. Computer Methods in Biomechanics and Biomedical Engineering. 2016;**19**(13):1443-1455. DOI: 10.1080/10255842.2016.1149575

[41] Dong J, Sun Z, Inthavong K, Tu J. Fluid–structure interaction analysis of the left coronary artery with variable angulation. Computer Methods in Biomechanics and Biomedical Engineering. 2015;**18**(14):1500-1508. DOI: 10.1080/10255842.2014.921682

The Influence of a Diamagnetic Copper Induced Field on Ion Flow and the Bernoulli Effect in Biological Systems

Marcy C. Purnell

Abstract

The Bernoulli Effect describes the principle of conservation of energy that optimizes pressure and motion in fluid flow and may be applied to fluid dynamics in vascular arterial and cellular membrane flow. One mechanism that is known to influence fluid flow that has *not* been included in the Bernoulli Effect equations is viscosity or resistance to flow. To date the liquid phase of matter with regards to the relationship between viscosity, pressure and flow is the least well understood of all phases. Recent cellular studies suggest that a diamagnetic copper influenced dielectrophoretic electromagnetic field may induce a Bernoulli Effect within biological systems. The data presented here suggests that an increased viscosity via this copper influenced dielectrophoretic electromagnetic field may significantly contribute to this Bernoulli Effect or conservation of energy while positively impacting cellular health and function via both kinetic and potential bio-energy influences in biological systems.

Keywords: Bernoulli effect, copper, diamagnetism, kinetic bio-energy, potential bio-energy, viscosity

1. Introduction

The Bernoulli Effect was formulated by a Swiss mathematician, Johann Bernoulli in 1738 to describe the principle of conservation of energy in fluid flow and can be applied to fluid dynamics in vascular arterial and cellular membrane flow [1–3]. The study of the Bernoulli Effect can enhance an understanding of how pressure relates to motion and energy to drive physiology in these areas of the body. One known mechanism for inducing flow that has *not* been included in the Bernoulli Effect equation is viscosity or resistance to flow [3]. To date, the liquid phase with regards to viscosity, pressure and flow is the least well understood of all the phases of matter [4, 5]. Viscosity can be defined as quantifying the internal frictional force that arises between adjacent layers of fluid that are in motion as can be found in plasma and cellular membrane flow [6].

Since water constitutes ~75 percent of the fluids that flow in the adult human body, water can be seen as not only critical to the sustenance of the physiological functions of life but also to the understanding of the conservation of energy in fluid flow of the body [7]. Studies of water have shown the potential of water-dielectric interfaces (as seen with chloride and water in plasma flow and cellular membranes) in electrostatic/potential energy harnessing and harvesting [8]. Recent cellular studies have also suggested that water and molecular attractions may play significant roles in the harnessing of energy components such as kinetic and potential bio-energy at the interface of cell membranes and in plasma flow [9–13]. Water that resides adjacent to hydrophilic surfaces/membranes appears to have defining characteristics that differ from bulk/free water (outside membranes and in the environment) and these unique features may correlate to the capacity to use magnetic attraction to harness energy and facilitate flow and movement of ionic solutions within plasma and across cell membranes [9, 14, 15]. Also, the use of a dielectrophoretic electromagnetic field (DEP EMF) that is generated with the influence of the noble diamagnetic metal, copper appears to have a significant impact on cellular function in biological systems [9, 14, 15]. Decades ago, research on diamagnetic copper and the role it plays in living systems began after the discovery that it was necessary for hemoglobin formation in rats, yet copper and its defining attributes of its contributary role in biological systems remain elusive to date [16]. Could viscosity that is not included in the Bernoulli equation by Johann Bernoulli be a significant component of the actual harnessing of magnetic energy in fluid flow in biological systems? It is known that a magnetorheological fluid becomes thicker and more viscous when subjected to a magnetic field [10]. The influence of diamagnetic copper on the generation of a.

DEP EMF appears to increase viscosity and change water structure in these kinematic viscosity and bubble coalescence studies presented below. This data suggests that this increase in viscosity may be related to a magnetic structural shift of the diamagnetic chloride's attraction to water and other materials in living systems [9]. This data, along with other recently published studies suggest that the influence of this diamagnetic metal (copper) induced DEP EMF on the dielectric anion, chloride may increase viscosity and decrease pressure in the flow in living systems thereby offering alternative kinetic and potential bio-energy sources (conservation of energy/ Bernoulli effect) for plasma and membrane flow in cellular functions within biological systems [9, 12, 13, 17]. Historically, increased viscosity in fluid flow in living systems (i.e., plasma flow) has not been desirable due to its association with stagnation and dysfunction of fluid flow. With the use of the data from the kinematic viscosity and bubble coalescence studies along with theory and computational equations, we will discuss some possible unknown characteristics of magnetism and viscosity and how it may impact conservation of energy in fluid flow in biological systems.

2. The role of diamagnetic copper in the generation of dielectrophoretic electromagnetic fields

Copper is an essential trace element that is vital to the health of all living things. While the importance of copper in health maintenance is widely accepted, exactly how this trace element functions within biological systems has been poorly defined to date. It is known that diamagnetic materials such as copper and chloride are repelled by and flow in *opposition* to a magnetic field. In contrast, paramagnetic materials such as sodium and ferromagnetic materials such as iron are attracted to and flow *with* the magnetic field [9]. Diamagnetic materials possess complete shells which behave as electric current loops that orient themselves in specific ways in magnetic fields. Copper is diamagnetic because unpaired electrons in the 4 s orbitals are localized to form metallic bonds. Historically, it has been thought that diamagnetism offers a weak or negligible contribution to a magnetic field. However, recent data suggest that diamagnetic metals such as copper may indeed play a significant role in the energy of life or the internal energy components of kinetic and potential bio-energy by the facilitation of a separation of charge. Charge separation between the position and negative ions is essential to facilitate movement in and around membranes and plasma cells. The zone within or near membranes is usually fixed and negative while the positive charges by contrast are free to diffuse in the regions around and beyond [14]. The persistence of positive and negative attraction that occurs in lower magnetic states (weaker field flow) appears to require a phosphoryla-tion of ATP to provide the energy to facilitate charge separation required to induce flow and maintenance of ion differentials in and around cells. Due to the nature of the need for charge separation (via an energy source) within an electromagnetic field to facilitate movement of positive charges (cations such as potassium, sodium, magnesium, hydrogen and calcium) across membranes and in the plasma, the addi-tion of a diamagnetic metal to the generation of this electromagnetic field in order to create a DEP EMF is essential (**Figure 1**). A DEP EMF induces dielectrophoresis or a phenomenon in which a force is exerted on dielectric particles such as chloride and changes the magnetic attraction from positive and negative attract to *like attracts like* or *like likes like* (**Figures 1** and **2**) [9, 14, 15]. Data suggest that this magnetic shift may actually represent a change in viscosity and a harnessing of kinetic and potential bio-energy in living systems.

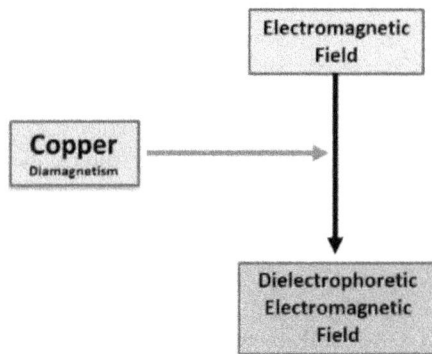

Figure 1.
The addition of the noble diamagnetic metal, copper to essential to generate a dielectrophoretic electromagnetic field. Copper can be said to perform as a field separator between cations and anions that generates a magnetic shift in attraction from positive and negative attract to "like attracts like" [9, 14].

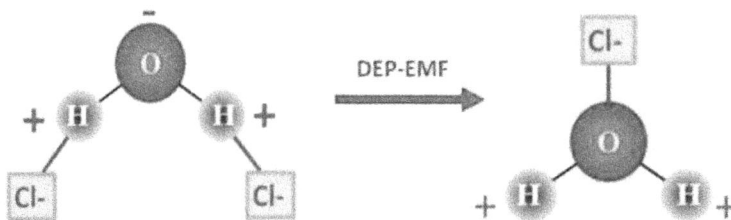

Figure 2.
A polar water molecule in low and high magnetic states: With decreased magnetic energy that may occur outside a living organism or in free waters in the environment, where positive and negative charges are known to attract [9, 14], the negatively charged chloride anion may be attracted to the positively charged hydrogen side of the polar water molecule. This water form may be seen in "unstructured/free water." with increased magnetism and internal potential energy that may reside at the membranes of and within living organisms [9, 14], the negatively charged chloride anion is attracted to the negatively charged oxygen side of the polar water molecule (structural change). Positive and negative charges are no longer attracted since there is a magnetic shift to "like attracts like" [9] or "like likes like" [14].

3. Dielectrophoretic electromagnetic field induced effects on kinematic viscosity

Dielectrophoresis (DEP) has been known to influence the flow and movement of microparticles, nanoparticles and cells [18, 19]. DEP can be explained as the net force encountered by a dielectric (polarized) particle in an electric field [20]. This force is impactful on all charged and uncharged particles and all particles exhibit dielectrophoretic activity in the presence of non-uniform electric fields. The strength of the force is dependent on the medium, electrical properties of the particles, the size and shape of the particles and the designed frequency of the field. This DEP force (F_{DEP}) can be written where E is the electric field and m(ω) is the induced dipole moment on the particle (Eq. (1)) [9]:

$$F_{DEP} = \left[m(\omega) \bullet \nabla \right] E \tag{1}$$

DEP can influence a polarizable particle (ion) that is suspended in a medium that is driven by alternating current (ac) or direct current (dc). When a particle that is more polarizable (positively charged) than the surrounding medium the net movement of the particle is oriented towards the region of the highest field flow/strength or positive dielectrophoresis (pDEP). Conversely, particles with polarizability less than that of the medium move towards the region of the lowest field gradient (in opposition to the field) or negative dielectrophoresis (nDEP). The chloride anion is a diamagnetic ion with possible dielectric properties and is therefore repelled by a magnetic field and orients in opposition to the field causing a repulsive force. The positively charged cations of sodium, potassium, magnesium, calcium etc., appear to follow the flow of the field.

In lower magnetic states, positive and negative charges are known to attract thereby allowing chloride anions to form hydrogen-bonded bridges with water molecules while the cations (sodium, potassium, calcium, magnesium etc.) bond to the oxygen side of the water (**Figure 2**) [9, 21]. In a higher magnetic state that may occur in the presence of a DEP EMF, there appears to be a shift in magnetic attraction that has been termed as "*like attracts like*" or "*like likes like*" (**Figure 2**) [9, 14, 15]. This magnetic attraction shift may cause the chloride anion to maintain a different orientation to the water molecule allowing for covalent (stronger) bonding between the chloride and oxygen (i.e., biochloride) as well as between the hydrogen molecules [9, 14, 15] (**Figures 2 and 3**). This magnetic restructuring may coincide with and manifest in micrographs as changes in the bubble coalescence (**Figure 4**). Dr. Gerald Pollack is a pioneer in the science of water structure and he refers to the bubble or droplets noted in the water coalescence studies as vesicles. He hypothesizes that these vesicles change characteristics depending on the phase of water inside (that exists as water vapor) that is generated from the energy they absorb [22]. Dr. Pollack has also identified the phenomenon of EZ water where he discusses how water at the membranes and inside a cell is structured differently than in free/bulk water in nature. He refers to this structured or EZ water as the fourth phase of water [22]. The magnetic shift or change in attraction to "*like likes like*" appears to create an exclusion zone (EZ water) adjacent to the membrane that is a negatively charged crystalline structure or what we have termed biochloride (BCl−) while the cations continue to reside in the free/bulk water zone (**Figure 5**) [22]. While ion differential flows across the membranes in most cells, the red blood cell is a torus and carries the differential on the surface of this torus thereby facilitating plasma flow [12, 23]. The diamagnetic chloride is known to play a significant role in the flow of Band3/AE1 anion exchange while the cations reside outside the negative membrane surface in the Stern layer [12, 23].

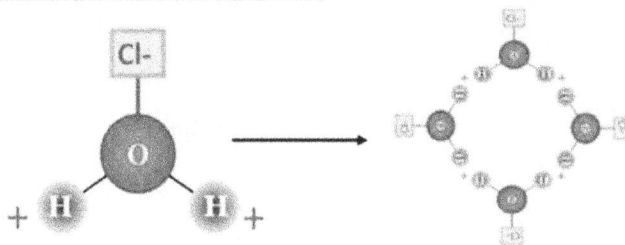

Figure 3.
Like attracts like - covalent bonding. In the presence of a non-uniform DEP EMF, there is a "like attracts like" or "like likes like" shift in magnetism. The chloride anion forms more stable covalent bonding (versus less stable hydrogen bonding) with the oxygen side of the polar water molecule and the hydrogen atoms do not form hydrogen bonding between other molecules, but instead form more stable covalent bonds between each other.

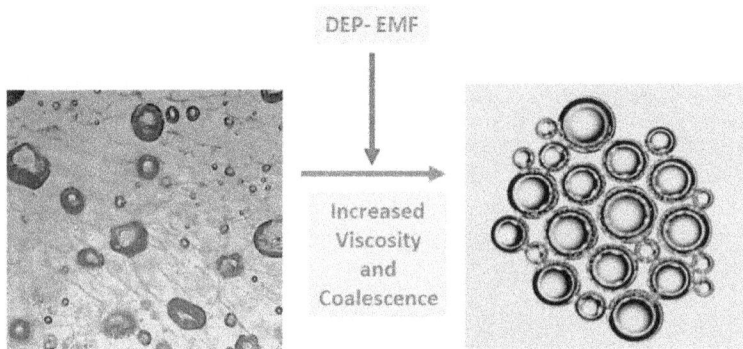

Figure 4.
*Water structure studies in the presence of a dielectrophoretic electromagnetic field (for 30 minutes). A 20°F hypotonic saline solution (left) was examined under a microscope 40X (left) with no exposure to dielectrophoretic electromagnetic field and after a 30-minute exposure (right) to the DEP EMF, notice the change in the coalescence of the bubbles (vesicles) that occurs along with the increased viscosity (**Table 1**). The bubbles/vesicles change form as they absorb energy as well as the phase of water (water vapor) inside the vesicles [22].*

In addition to these magnetically driven structural changes, this data suggests that the application of a non-uniform 2.5 ampere DEP EMF may significantly increase kinematic viscosity (resistance to flow) (**Table 1**). Kinetic viscosity (v or "nu") is the ratio of the viscosity of a fluid to its density (η/ϱ) or a measure of the resistive flow of a fluid under the influence of gravity (Eq. (2)) [24].

$$v = \eta / n \qquad (2)$$

A common unit that is used for kinematic viscosity is the square centimeter per second (cm^2/s) or Stokes (St) named after the Irish mathematician and physicist George Stokes. In our kinematic viscosity studies, using transparent plastic tubes with containing a chrome steel ball and a slower teflon ball, we found a significant increase in the hypotonic saline solution's kinematic viscosity that had been exposed to the non-uniform 2.5 ampere DEP EMF and the control saline solution that had not been exposed to the DEP EMF (Control Mean 8.29 cm^2/s; Treated Mean 7.08 cm^2/s; p = 0.001) (**Table 1**).

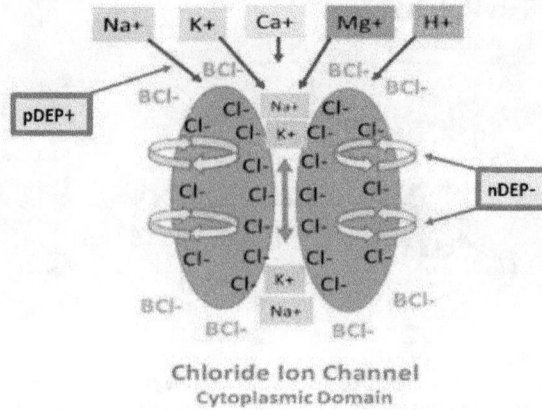

Chloride Ion Channel
Cytoplasmic Domain

Figure 5.
*Diamagnetic anisotropy is seen here where the diamagnetic chloride anions (Cl-) spin is in opposition to the flow of the field (**nDEP-**) but are driven/fueled by the potential (molecular ionic attraction potential) energy. The positively charged cations (Na⁺) (K⁺) (Mg⁺) (Ca⁺) (H⁺) flow with the field (**pDEP+**). Also, the biochloride (**BCl-**) and EZ (4th phase) water reside at both the plasma and cytoplasmic domains at the membrane and allow for the cations to enter the membrane through this magnetic shift in attraction. The magnetic shift to "like attracts like" allows for free flow of cations through the membranes (kinetic energy) possibly without a need for phosphorylation of ATP, thereby utilizing magnetic energy harnessed within the EZ-structured water/bio-chloride. This may offer a conservation of energy (Bernoulli effect) mechanism by using magnetism instead of ATP phosphorylation to drive ion differential maintenance. Note: The ionic differential of a red blood cell in the plasma is on the surface of the torus versus across the membrane [12, 23].*

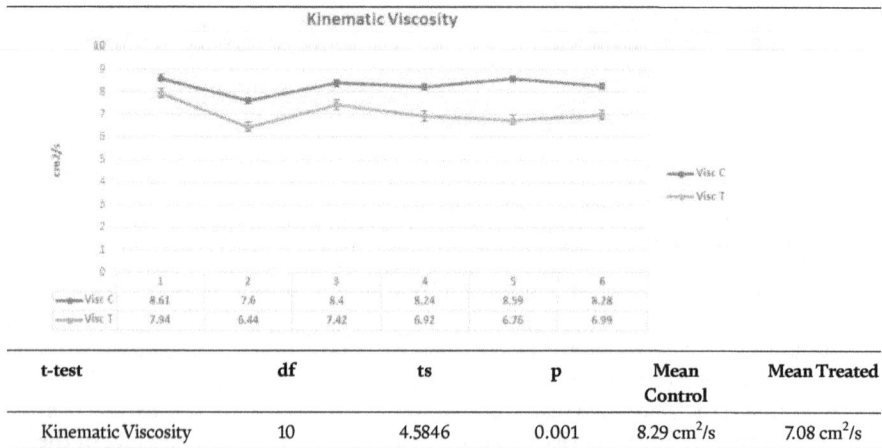

	1	2	3	4	5	6
Visc C	8.61	7.6	8.4	8.24	8.59	8.28
Visc T	7.94	6.44	7.42	6.92	6.76	6.99

t-test	df	ts	p	Mean Control	Mean Treated
Kinematic Viscosity	10	4.5846	0.001	8.29 cm²/s	7.08 cm²/s

Table 1.
Kinematic viscosity study- t-tests of viscosity (cm²/s) between DEP EMF control versus treated hypotonic saline solution.

4. Kinematic viscosity induced effect on pressure, kinetic bio-energy and potential bio-energy

One known effect of an increase in viscous forces is the ability to conduct negative work (drop in fluid pressure along the flow path) on the fluid, reducing its macroscopic mechanical energy while increasing the internal energy (microscopic kinetic/molecular ionic potential energy) and resulting in a slight increase in temperature [4]. Since pressure is a measure of fluid mechanical energy per unit volume, the correlation of a decrease in macroscopic mechanical pressure along the flow path (i.e., through vessels or across membranes) along with the increased internal energy (kinetic and potential bio-energy) of the fluid is noted in the Bernoulli Equation below (Eq. (3)) [25]:

$$P_1 + \tfrac{1}{2}\rho v_1^2 + \rho gh_1 = P_2 + \tfrac{1}{2}\rho v_2^2 + \rho gh_2 \qquad (3)$$

The variables P_1, v_1 and h_1 refer to the pressure, speed and height of the fluid at the initiation point and the variables P_2, v_2^2 and h_2 refer to the pressure, speed and height of the fluid as it flows to another point. Also, $\tfrac{1}{2}\rho v^2$ = kinetic energy per unit volume (Eq. (4)) and ρgh = potential energy per unit volume [25]. It is known that $P_2 < P_1$ for as the fluid transitions along the flow path, the pressure energy decreases while the kinetic energy and potential energy increase. In other words, increased fluid speed creates decreased internal pressure. This Bernoulli Equation as well as viscosity induced by a DEP EMF may indeed correlate to a conservation of energy principle where the increase in viscosity may trigger the lowering of pressure in the regions where the flow/velocity is increased along with increased kinetic bio-energy and potential bio-energy.

The average kinetic energy (KE_{avg}) per unit volume (V) of flowing fluid can be expressed in terms of the fluid density (ρ) and maximum flow velocity (v_m) (Eq. (4)) [25]:

$$KE_{avg} / V = (\tfrac{1}{2}\rho)\left(v_m^2 / 3\right) \qquad (4)$$

When the kinetic energy of fluid is examined with regards to laminar flow (as occurs in the plasma and across the cellular membranes) one must take the average of the velocity (shear stress and velocity gradient = viscosity) squared into account. The relationship between velocity and viscosity again speaks to the strong correlation viscosity has to kinetic energy. Increased viscosity and increased velocity may indeed be a significant factor in the internal kinetic energy (and pressure changes as well) and temperature regulation in biological systems. Might this non-uniform 2.5 ampere DEP EMF driven increase in kinematic viscosity offer a Bernoulli effect/conservation of energy (via increased kinematic viscosity→ increased averaged velocity → decreased pressure) in biological systems? Could this also have implications for membrane flow, plasma flow, blood pressure regulation and temperature control in the living organisms [9, 12, 23]?

One area where potential energy of a moving fluid in biological systems may reside would be in or near the membranes of cells. According to Dr. Pollack and EZ water (fourth phase) can be seen to function as a battery [22]. There have also been additional studies that suggest this structured or EZ water may increase the magnetic property or diamagnetism of the chloride ion and significantly modulate chloride ion channels both across the membranes and on the surface of red blood cells [9, 11–13, 23]. As stated earlier, chloride is a diamagnetic ion (that displays dielectric properties) that is both repelled and driven by the magnetic field. This dielectric behavior may offer a diamagnetic anisotropy mechanism in the membrane (**Figure 6**). *In vitro* and human studies using a copper influenced DEP EMF have shown a significant increase in chloride channel modulation [9, 11–13, 23]. The paramagnetic cations (Na^+, Mg^+, Ca+, K^+, H^+) move in the direction of the field or with the flow of the field, while the diamagnetic chloride anion will move in opposition to the field, acting as a field separator and facilitator of movement (through repulsion) for the cations (**Figure 6**). Chloride ion channel inhibition has been

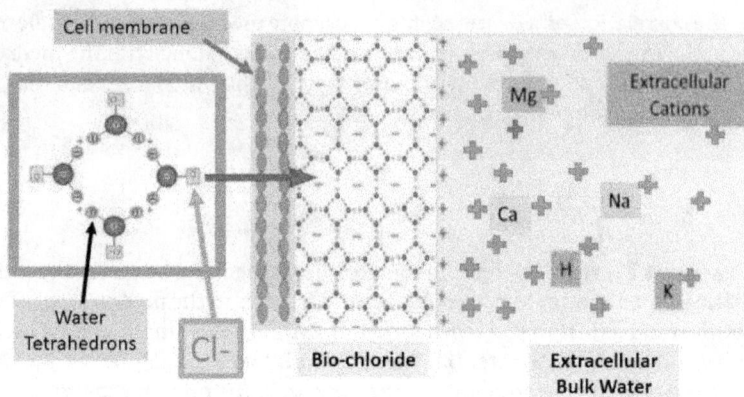

Figure 6.
*The internal potential molecular ionic attraction energy forms an exclusion zone or the fourth phase of water. The internal potential energy driven molecular ionic attraction facilitates a "like attracts like" or "like likes like" (EZ water) where chloride and water create crystalline structure with water tetrahedrons as they absorb magnetic energy [9, 14, 22] (**Figures 2–4**). The hydrogen ions bond or coalesce (molecular ionic attraction potential energy) and the negatively charged diamagnetic chloride anion (with an oxygen operating as a charge separator- bio-chloride/EZ water) spins in opposition to the field facilitating the passage (kinetic energy) of the charged cations through the field/membranes. This constitutes diamagnetic anisotropy and basically catapults the cations through the membrane with electromagnetically driven kinetic energy and potential energy in biological systems (**Figure 5**) [9, 14, 22].*

noted with increased levels of extracellular ATP [26]. Volume sensitive chloride channels have also been found to be regulated by intracellular ATP concentrations [27]. Since the application of this DEP EMF has shown significant upregulation of chloride ion channels in *in vitro* studies, might this external DEP EMF application offer an external energy source that is not dependent or regulated by ATP levels in the organism? Could this offer an energy conservation source for cellular function within biological systems?

5. Dielectrophoretic electromagnetic field induced effects on vesicle coalescence

Five millimeters of a 3 mM hypotonic saline solution prepared with laboratory-grade deionized water and molecular biology grade NaCl (Promega, Madison, WI) were placed on a glass slide in a 20°F freezer for 30 minutes. This same hypotonic saline solution was then exposed to a non-uniform 2.5 ampere DEP EMF field using a compilation of 6 stainless steel rings around a center copper ring for 30 minutes and five millimeters of this DEP EMF treated solution was also placed on a slide in a 20°F freezer for 30 minutes. The control and treated slides were then examined under 40x light microscopy and micrographs were immediately taken. Upon analysis of the micrographs, there were noticeable differences in the bubble/vesicle coalescence between the frozen hypotonic saline solution that was not exposed to the DEP EMF and the frozen hypotonic solution that was exposed to the DEP EMF (**Figure 4**). The effects of viscosity on bubble/vesicle coalescence have been studied both experimentally and numerically and a higher viscosity in liquids showed an increase in coalescence time and characteristics and when compared to lower viscosity liquids [22, 28, 29]. Upon analysis of the micrographs, there were noticeable differences in the bubble/vesicle coalescence between the frozen hypotonic saline solution that was not exposed to the DEP EMF and the frozen hypotonic solution that was exposed to the DEP EMF (**Figure 4**). This characteristic change in vesicle organization in the presence of the DEP EMF may correspond to a change in how water and ions orient to each other (**Figures 2** and **3**).

6. Conclusions

Life sustenance is an energy consuming process where biological systems must transform energy from one form to another through complex chains of biochemical and physiochemical events. Historically, increased viscosity of plasma and fluid flow in the body has been associated with increased coagulation, dehydration and other potentially unwanted pathophysiology. This data suggest that an *externally* applied copper influenced non-uniform DEP EMF may increase viscosity due to a magnetic restructuring of water and ions (via dielectrophoresis) thereby harnessing kinetic bio-energy (plasma and membrane flow) and potential bio-energy (EZ water-biochloride, diamagnetic anisotropy) in biological systems. This harnessing of energy may add an additional energy source that is available to cells, thereby conserving the need for higher levels of *internally* generated energy such as ATP for phosphorylation. Since cell stress/inflammation, the basis of all disease, is due to lack of adequate energy sources to carry out cellular physiological function, the ability to decrease cell stress by providing an externally applied (i.e., battery charge) energy resource for the cell to use and relieve energy deficits may open the door for new areas of research with multiple morbidities. The application of a copper influenced dielectrophoretic electromagnetic field appears to increase viscosity (harnessing of magnetic energy), decrease pressure (via decreased macroscopic mechanical energy) while conserving energy and facilitating motion/fluid flow in biological systems, i.e., the Bernoulli Effect (**Figure 7**) [9, 11–14, 22, 23, 30–34]. Therefore, this field application and the effect on ion flow has the potential to address multiple co-morbidities such as: hypertension, cancer, wound care, organ dysfunction, infectious disease and cardiovascular disease [9, 12, 13, 17, 23, 33, 34].

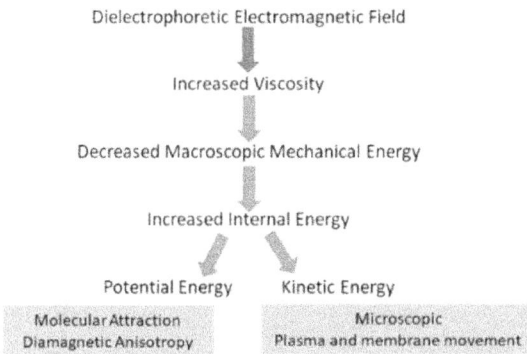

Dielectrophoretic Electromagnetic Field

Increased Viscosity

Decreased Macroscopic Mechanical Energy

Increased Internal Energy

Potential Energy   Kinetic Energy

Molecular Attraction   Microscopic
Diamagnetic Anisotropy   Plasma and membrane movement

Figure 7.
Dielectrophoretic electromagnetic Fields's generation of potential and kinetic bio-energy and possible conservation of energy (a Bernoulli effect in biological systems) where internal, kinetic and potential energy remain constant or increase in exchange for a decrease in system pressure and an increase in temperature (homeostasis of ion differentials, blood pressure and temperature).

Author details

Marcy C. Purnell
School of Nursing, Louisiana State University Health Science Center,
New Orleans, Louisiana, United States

*Address all correspondence to: mpurn1@lsuhsc.edu

IntechOpen

References

[1] Lawrence-Brown MMD, Liffman K, Semmens JB, & Sutalo ID. Vascular arterial Haemodynamics. In: Fitridge R, Thompson M, editors. Mechanisms of Vascular Disease: A Reference Book for Vascular Specialists. Adelaide: University of Adelaide Press. 2011. 8

[2] Yu H, Lin Z, Xu L, Liu D, Shen Y. Theoretical study of microbubble dynamics in sonoporation. Ultrasonics. 2015; 61, 136-144

[3] Vogel S. Organisms that capture currents. Scientific American. 1978; 239 (2): 128-139

[4] Schmelzer JWP. Pressure dependence of viscosity. J Chem Phys. 2005; 122, 074511

[5] McMillan PF, Wilding MC. High pressure effects on liquid viscosity and glass transition behaviour, polymorphic phase transitions and structural properties of glasses and liquids. Journal of Non-Crystalline Solids. 2009; 355, 722-732

[6] Viswanath DS, Ghosh TK, Prasad DL, Dutt NVK, Rani KY. Viscosity of Liquids: Theory, Estimation, Experiment, and Data. 2007. Springer. Netherlands

[7] Kuriyan R. Body composition techniques. Indian J Med Res. 2018 Nov;148(5): 648-658

[8] Cherepanov DA, Feniouk BA, Junge W, Mulkidjanian AY. Low dielectric permittivity of water at the membrane interface: Effect on energy coupling mechanism in biological membranes. Biophysical journal. 2003;85(2): 1307-1316

[9] Purnell, M. & Skrinjar T. The Dielectrophoretic dissociation of chloride ions and the influence on diamagnetic anisotropy in cell membranes. Discovery Medicine. 2016a; 22 (122): 257-273

[10] Yu J, Ma E, & Ma T. Harvesting energy from low-frequency excitations through alternate contacts between water and two dielectric materials. Scientific Reports. 2017; 7: 17145

[11] Purnell, M. Bio-electric field enhancement: The influence on hyaluronan mediated motility receptors in human breast carcinoma. Discovery Medicine. 2017; 23 (127): 259-267

[12] Purnell M, Butawan MA, & Ramsey RD. Bio-field array: A dielectrophoretic electromagnetic toroidal excitation to restore and maintain the golden ration in human erythrocytes. Physiological Reports. 2018; 6 (11), e13722

[13] Purnell M, Butawan MBA, Bingol K, Tolley EA, & Whitt MA. Modulation of endoplasmic reticulum stress and the unfolded protein response in cancerous and noncancerous cells. SAGE Open Med. 2018; 20 (6): 205012118783412

[14] Pollack G, Figueroa X, & Zhao Q. Molecules, water and radiant energy: New clues for the origin of life. International Journal of Molecular Sciences. 2009; 10 (4): 1419-1429

[15] Ball P. Like attracts like. Nature. 2012; Retrieved from: https://www.nature.com/news/like-attracts-like-1.10698

[16] Hart EB, Steenbock H, Waddell J, Elvehjem CA. Iron in nutrition. VII. Copper as a supplement to iron for hemoglobin building in the rat. 1928. J Biol Chem. 2002; 277(34): e22. PMID: 12243126

[17] Purnell, M. Bio-Field Array: The influence of JMY expression on cytoskeletal filament behavior during apoptosis in human triple negative breast Cancer. Breast Cancer: Basic and Clinical Research. 2019; https://doi.org/10.1177/1178223419830981

[18] Wissner-Gross AD. Dielectrophoretic reconfiguration of nanowire interconnects. Nanotechnology. 2006; 17:4989-4990

[19] Pommer MS, Zhang Y, Keerthi N, Chen D, Thomson JA, Meinhart H, Soh T. Dielectrophoretic separation of platelets from diluted whole blood in microfluidic channels. Electrophoresis. 2008; 29(6): 1213-1218

[20] Lungu M, Neculae A, Bunoiu M, Strambeanu N. considerations on the nanoparticles manipulation in fluid media using dielectrophoresis. Rom J Phys. 2011; 56: 749-756

[21] Mancinelli R, Botti A, Bruni F, Ricci MA, Soper AK. Hydration of sodium, potassium, and chloride ions in solution and the concept of structure maker/breaker. J Phys Chem B. 2007; 111(48): 13570-135777

[22] Pollack G. The fourth phase of water. EDGESCIENCE. 2013; 16: 14-18

[23] Purnell M, & Ramsey RD. The influence of the Golden ratio on the erythrocyte, Erythrocyte, Anil Tombak, IntechOpen. 2019; Doi: 10.5772/intechopen.83682

[24] Elert G. Viscosity-The Physics Hypertextbook, Physics info. Brooklyn, N.Y. 2010. retrieved from https://physics.info/viscosity

[25] Nave CR. Hyperphysics. 2000. Retrieved from: http://hyperphysics/phy-astro.gsu.edu/hbase/hframe.html

[26] Voss, A.A. Extracellular ATP inhibits chloride channels in mature mammalian skeletal muscle by activating $P2Y_1$ receptors. J Physiol. 2009; 587 (23): 5739-5732

[27] Hilgemann DW. Cytoplasmic ATP-dependent regulation of ion transporters and channels: Mechanisms and messengers. Annu Rev Physiol. 1997; 59, 193-200

[28] Orvalho, S., Ruzicka, M.C., Olivieri, G., & Marzocchella, A. Bubble coalescence: Effect of bubble approach velocity and liquid viscosity. Chemical Engineering Science. 2015; 134: 205-216

[29] Sanada T, Watanabe M, & Fukano T. Effects of viscosity on coalescence of a bubble upon impact with a free surface. Chemical Engineering Science. 2005; 60(19): 5372-5384

[30] Badeer, H.S. Hemodynamics for medical students. Adv Physiol Educ. 2001; 25: 44-52

[31] Wang, Y., Steele, C.R., & Puria, S. Cochlear outer-hair-cell power generation and viscous fluid loss. Sci Rep. 2016; 6, 19475; doi: 10.1038/srep19475

[32] Rao, S.G., Patel, N.J., & Singh, H. Intracellular chloride channels: Novel biomarkers in disease. Front Physiol. 2020; Retrieved from: https://doi.org/10.3389/fphys.2020.00096

[33] Purnell, M & Whitt, M. Bioelectrodynamics: A new patient care strategy for nursing health and wellness, Holistic Nursing Practice. 2016; 30: 4-9

[34] Purnell, M. & Skrinjar, T. Bioelectric field enhancement: The influence on membrane potential and cell migration In vitro. Advances in Wound Care. 2016b; 5 (12): 539-545

Analysis of Geometric Parameters of the Nozzle Orifice on Cavitating Flow and Entropy Production in a Diesel Injector

Fraj Echouchene and Hafedh Belmabrouk

Abstract

In this chapter, we investigated the effect of geometric parameters of the nozzle orifice on cavitating flow and entropy production in a diesel injector. Firstly, we analyzed the effect of some parameters of diesel injector such as the nozzle length and the lip rounding on cavitating flow. In the second parts, we studied the entropy production inside the diesel injector in several cases: -single phase and laminar flow,- single phase and turbulent flow and –tubulent cavitating flow. In the last case, the mixture model cupled with k-ε turbulent model has been adopted. The effects of average inlet velocity and cavitation number on entropy production have been presented and discussed. The results obtained show that the discharge coefficient is weakly influenced by the length of the orifice and the radius of the wedge has a large effect on the intensity and distribution of cavitation along the injection nozzle. On the other hand, the study of entropy production inside the diesel injector shows that the entropy production is important near the wall and increases whith increasing the average inlet velocity and pressure injection.

Keywords: Cavitation, Diesel injector, Entropy generation, Simulation, Mixture model

1. Introduction

The fuel flow through injector nozzles affects the spray formation, the atomization phenomenon of the liquid fuel and, therefore, the efficiency of the combustion process and pollutant emission. Modern passenger cars and trucks use higher injection pressures than early models to improve the atomization of fuel in order to reduce soot emission of internal combustion engines. Diesel engine injectors often operate at injection pressures about 2000 bar. The high injection pressure and the abrupt change of the orifice section of the injector allows to have a pressure drop below the saturated vapor pressure and consequently the development of cavitation. Cavitation has a great effect on both the fuel injection process and the performance of an engine. Cavitation generated at the entrance of the orifice affects the fluid flow and the atomization of the injected liquid jet [1–3]. Cavitation is often observed in pumps, inducers, hydraulic turbines, propellers, fuel injectors, and other fluid devices [4, 5]. However, cavitation has positive effects in some biomedical and industrial applications such as shock wave lithotripsy, water disinfection and organic compounds decomposition, etc. [6, 7].

The study investigation of cavitation phenomenon in the injection orifice is useful even important to control and optimize the atomization process.

The high injection pressure (> 2000 bar), the high speed flow [8] and the small dimensions of the injection nozzle make the studies experimental. In addition, experiments were performed on large-scale and transparent injector configurations to visualize the phenomenon of cavitation [3, 9–15].

Confronted with experimental difficulties, several theoretical and numerical studies have been developed to study this cavitation problem in a real diesel injector [16–22]. In a previous study [17], we studied numerically, using the mixing model, the effect of the wall roughness of the orifice injection on the cavitation phenomenon. In another work [23], we studied the effect of inlet corner radius of orifice injection on the flow characteristic and the development of cavitation. We noticed a reduction in the intensity of cavitation when corner radius increases.

The relative risk of erosion of the inner wall of the diesel injector orifice due to cavitation has been studied by [16, 22]. Xue et al. studied the effect of cavitation in a multi-hole injector on the transient flow characteristic in a 3D asymmetric configuration using a two-phase (liquid–vapor) model [21]. The effect of the needle lift was analyzed by these authors. They showed a difference in velocity profile and cavitation within the holes.

Torelli et al. [24] performed a 3D simulation in a five-hole diesel mini-injector to model the internal flow of the nozzle using three types of fuel (full-range naphtha, light naphtha and n-Dodecane). They have show that the cavitation is strongly related to the saturating vapor pressure of different fuel.

In this chapter, we aim to investigate the cavitating flow inside a Diesel injector using the mixture model and taking into account the turbulence. A parametric analysis of the size and the shape of the injector is carried out. The entropy production inside the diesel injector in several cases: -single phase and laminar flow,- single phase and turbulent flow and –tubulent cavitating flow is analyzed. Furthermore, the flow is simulated in the steady state as well as in the unsteady state.

2. Theoretical model

In this mixture model, the fluid (fuel) is composed of three phases: liquid, vapor and non-condensable gases (CO) which the mixture density is given by [25].

$$\rho_m = \alpha_v\rho_v + \alpha_g\rho_g + (1 - \alpha_v - \alpha_g)\rho_l \tag{1}$$

where ρ is the density and α is the volume fraction. The indices l, v and g denote the liquid, vapor and gas phases, respectively.

The mass fraction f_i can be calculated from this equation

$$f_i = \frac{\alpha_i \times \rho_i}{\rho_m} \tag{2}$$

The transport equations describing the cavitating flow inside the diesel injector are:

• Navier Stokes equations for the mixture

• k-ε turbulence model

• Transport equation of vapor fraction

2.1 Multiphase model

In this work, the following hypotheses are adopted:

- The mixture is assumed to be single-phase;

- The flow is assumed to be isothermal and incompressible;

- The fluid is Newtonian;

- The gravity force is neglected

The mass conservation equation of the mixture flow is [17]:

$$\frac{\partial}{\partial t}(\rho_m) + \nabla \cdot \left(\rho_m \vec{U}_m\right) = 0 \qquad (3)$$

where $\vec{U}_m$ present the velocity of the mixture.
The mixture momentum conservation equation is [17]:

$$\frac{\partial}{\partial t}\left(\rho_m \vec{U}_m\right) + \nabla \cdot \left(\rho_m \vec{U}_m \otimes \vec{U}_m\right) = -\nabla p + \nabla \cdot \left[(\mu_t + \mu_m)\left(\nabla \vec{U}_m + \left(\nabla \vec{U}_m\right)^T\right)\right] \qquad (4)$$

where μ_m is the laminar viscosity of the mixture and $\mu_t = \rho_m C_\mu \frac{k^2}{\varepsilon}$ is the turbulent viscosity. C_μ = 0.09, k is turbulent kinetic energy, ε is the dissipation rate.

It should be noted that the liquid and the vapor have the same velocity $\vec{U}_m$. Since the liquid is incompressible, we obtain $div\left(\vec{U}_m\right) = 0$. On the other hand, for the steady flow, the continuity equation becomes $div\left(\rho_m \vec{U}_m\right) = 0$.

2.2 k-ε turbulence model

Several experimental investigations have shown that turbulence has a significant effect on cavitating flows (e.g. [26]). Also, [27] studied the sensitivity of the cavitating flows to turbulent fluctuations. For the present computations, we use the standard k-ε turbulence [17, 23]:

$$\frac{\partial}{\partial t}\left(\rho_m k \vec{U}_m\right) + \nabla \cdot \left(\rho_m k \vec{U}_m\right) = \nabla \cdot \left[\left(\mu + \frac{\mu_t}{\sigma_k}\right)\nabla k\right] + P - \rho \varepsilon \qquad (5)$$

$$\frac{\partial}{\partial t}\left(\rho_m \varepsilon \vec{U}_m\right) + \nabla \cdot \left(\rho_m \varepsilon \vec{U}_m\right) = \nabla \cdot \left[\left(\mu + \frac{\mu_t}{\sigma_\varepsilon}\right)\nabla \varepsilon\right] + C_{1\varepsilon}\frac{\varepsilon}{k}P - C_{2\varepsilon}\frac{\varepsilon^2}{k} \qquad (6)$$

where P is the production term of turbulent kinetic energy given by:

$$P = \mu_t\left[(\nabla U_m) + (\nabla U_m)^T\right]\nabla U_m \qquad (7)$$

The standard values of the constants are: σ_k = 1.0, σ_ε = 1.3, $C_{1\varepsilon}$ = 1.44 and $C_{2\varepsilon}$ = 1.92 [17].

2.3 Cavitation model

According to [17, 28], the differential equation describing the transport of the vapor fraction is given by

$$\frac{\partial}{\partial t}\left(\rho_m f_v\right) + \nabla \cdot \left(\rho_m f_v \vec{U}_v\right) = R_e - R_c \tag{8}$$

Here R_e and R_c denote the evaporation and condensation rates, respectively. The rates R_e and R_c depend on the static pressure and the velocity as well as on the fluid properties. They depend also on the pressure fluctuations generated by the turbulence as well as on the turbulent kinetic energy k.

The following relations are usually adopted to express the rates R_e and R_c [29].

$$R_e = C_e \frac{\sqrt{k}}{\sigma} \rho_\ell \rho_v \left(\frac{2}{3}\frac{p_v - p}{\rho_l}\right)^{1/2} \left(1 - f_v - f_g\right) \qquad p \leq p_v \tag{9}$$

$$R_c = C_c \frac{\sqrt{k}}{\sigma} \rho_\ell \rho_v \left(\frac{2}{3}\frac{p - p_v}{\rho_l}\right)^{1/2} f_v \qquad p > p_v, \tag{10}$$

where $C_e = 0.02$ and $C_c = 0.01$ are calibration constants, f_v and f_g are vapor mass fraction and non-condensable gas mass fraction. The phase-change threshold pressure p_v is estimated from the following equation [17]:

$$p_v = p_{sat} + \frac{p_t}{2} \tag{11}$$

where p_{sat} present the vapor saturation pressure and $p_t = 0.39\rho_m k$ is the turbulence pressure [17]. The above relations show that the evaporation and condensation happen at the vapor pressure p_v and not at the saturation pressure p_{sat} as in laminar flows. The diffusion phenomenon between phases is neglected.

2.4 Entropy production

Taking into account the following assumptions:

- The slip velocity between the two phases is negligible;

- The diffusion of species through the interface is ignored

The entropy production rate for the mixture is then written [30]:

$$\bar{P}_{sm} = \frac{1}{T}\left(\Phi_{\mu m} + \Phi_{tm} + \Phi_{Dm}\right) \tag{12}$$

where $\Phi_{\mu m}$ is the average entropy production for the mixture, Φ_{tm} is the entropy generation due to turbulent shear stress and Φ_{Dm} is the entropy production due to the turbulent dissipation term.

The mean of the entropy production and the entropy production due to the turbulent shear stress can be written as follows [30]:

$$\Phi_{effm} = \Phi_{\mu m} + \Phi_{tm} = \mu_{effm}\left[\left(\nabla\cdot\langle\vec{U}_m\rangle\right) + \left(\nabla\cdot\langle\vec{U}_m\rangle\right)^{\mathrm{T}}\right] : \nabla\cdot\langle\vec{U}_m\rangle\right] \tag{13}$$

where $\mu_{effm} = (\mu_m + \mu_{t_m})$ is the effective viscosity of the mixture.

In a two-dimensional flow in an injection orifice, the entropy production in cylindrical coordinates (2D) is then written [30]:

$$\bar{P}_{sm} = \frac{\mu_{effm}}{T}\left[2\left(\left(\frac{\partial u_m}{\partial r}\right)^2 + \left(\frac{u_m}{r}\right)^2 + \left(\frac{\partial v_m}{\partial z}\right)^2\right) + \left(\frac{\partial u_m}{\partial z} + \frac{\partial v_m}{\partial r}\right)^2\right] + \frac{\rho_m \bar{\varepsilon}_m}{T} \quad (14)$$

3. Numerical method and nozzle geometry

To simulate the cavitating flow, the numerical code Fluent was used. This code is based on implicit finite volume scheme. The SIMPLE algorithm [31] is used for the pressure–velocity coupling. Grid generation process for performing finite volume simulations were carried out using GAMBIT (v2.3.16) program available with the commercial code Fluent. A first order implicit temporal discretization and a first order upwind differentiating scheme have been used. All under-relaxation factors range between to 0.2–0.4.

3.1 Nozzle geometry and boundary conditions.

Figure 1 illustrates the nozzle geometry of diesel injector in 2D axisymmetric. The geometric parameters of the nozzle are R_1 = 0.3 mm, R_2 = 0.1 mm and L_1 = 0.5 mm. The transition radius between inlet pipe and orifice is r_c.

In this study, a stationary single phase fluid is assumed as initial conditions. Uniform inlet and outlet static pressure were adopted as boundary conditions. A value of k_0 and ε_0 are imposed at the inlet.

3.2 Effect of the grid resolution

the mesh presents an essential parameter in fluid mechanics problems for the convergence of the solution. The existence of an important gradient of the physical (pressure, velocity and vapor fraction) requires a concrete study of the sensitivity of the mesh on the solution. The mesh sensitivity on the solution was studied under the following injection conditions: inlet pressure p_{in} = 5 bar and outlet pressure p_{out} = 1 bar.

The mesh grid is adapted several times (six times: see **Table 1**). In **Table 1**, n and m represents the number of meshes according to z and r axis in the orifice.

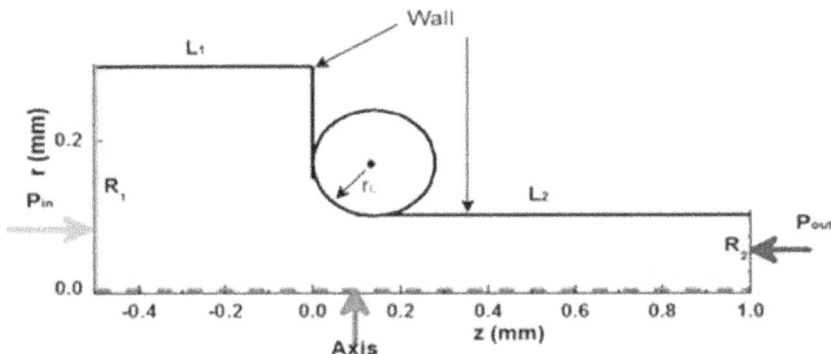

Figure 1.
2D-axisymmetric configuration of the diesel injector with boundary conditions.

Mesh	0	1	2	3	4	5
n × m	5x30	10x50	15x80	20x110	25x140	30x160

where n and m are the number of meshes according to z and r axis in the orifice.

Table 1.
Meshes tested in the simulation.

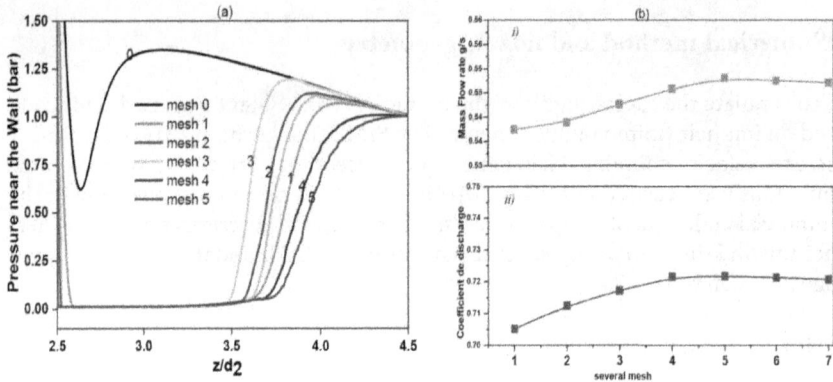

Figure 2.
Pressure profile near the wall using different mesh and b-mass flow rate and discharge coefficient for several grid resolutions.

The mesh effect on the local field (e.g the pressure field) and on the global coefficient (e.g discharge coefficient) along the wall of the nozzle is presented in **Figure 2(a,b)**. **Figure 2(a)**, indicates the pressure field distribution along the near wall at the orifice for several grids resolutions. It is clear that the local minimum pressure is strongly affected by the mesh size. The local pressure near the wall decrease with mesh size until it reaches a minimum value lower than the vapor saturation pressure.

In this region, the cavitation phenomenon appears. According to Raleigh-Plesset Equation [32], cavitation is governed by the local pressure. The effect of mesh on the masse flow rate and discharge coefficient is presented in **Figure 2b (i & ii)**. We notice a variation in the discharge coefficient C_d with the mesh. For example, C_d undergoes a variation of 1.6% going from 1 to 4 and then remains constant. Mesh N° 3 can be used for simulation.

4. Results and discussions

4.1 Steady flow

In this part, we study the effect of geometric parameters such as L_2/d_2, r_c/d_2 ratios and Reynolds number Re on the cavitation phenomenon in steady state.

4.1.1 Influence of the Nozzle length

The pressure drop in the injector is described by the discharge coefficient C_d that is the ratio of the effective mass flow rate to the theoretical maximum flow rate:

$$C_d = \frac{\dot{m}_{eff}}{\dot{m}_{ideal}} \tag{15}$$

where $\dot{m}_{eff} = \iint \rho_m \vec{v}_m . \overrightarrow{dA}$ is the effective mass flow rate and $\dot{m}_{ideal} = \pi R_2^2 \sqrt{2\rho_l (p_{in} - p_{out})}$ is the ideal flow rate.

Figure 3 shows the distributions of vapor volume fraction for several nozzle lengths (p_{in} = 10 bar and p_{out} = 1 bar). The latter (nozzle lengths) has a significant effect on the cavitation area. The cavitation region decreases when the length of nozzle increases.

We perceive a transition on the cavitation nature from fully developed cavitation (for $L_2 = d_2$ or $2d_2$) to incipient cavitation (for $L_2 > 2d_2$).

For a length $L_2 \leq 0.6$ mm, cavitation still takes place and occupies a large part of the orifice. Cavitation is highly developed in this situation. However for $L_2 \geq 0.6$ mm, the cavitation remains very confined. It is useful to also examine the effect of length at high injection pressures and taking into account the presence of the combustion chamber.

Figure 4, shows the discharge coefficient C_d as a function of L_2/d_2 ratio for a nozzle with $r_c/d_2 = 0$. it is clear that the discharge coefficient varies linearly with the L_2/d_2 ratio. When the L_2/d_2 ratio increases from 2 to 8, we notice a reduction of 13%

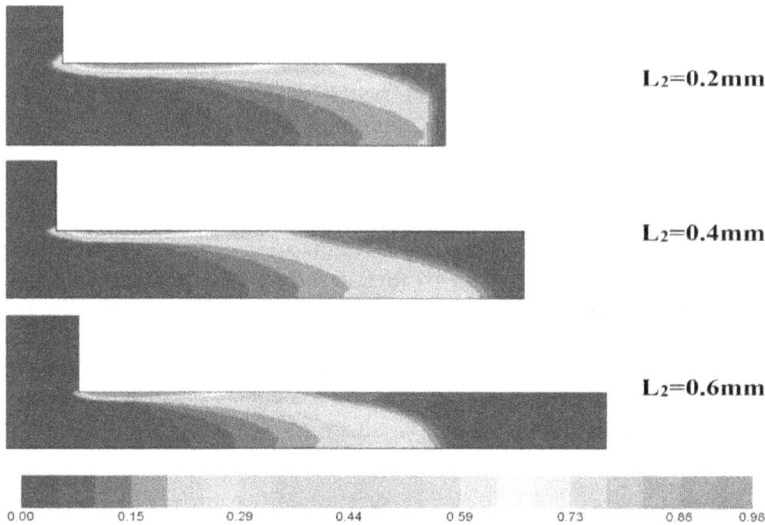

Figure 3.
Contour of vapor volume fraction for several nozzle lengths.

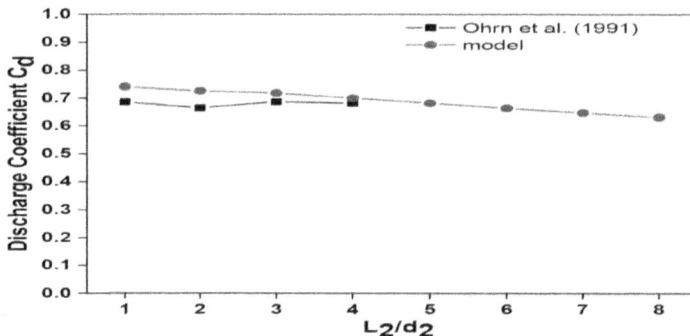

Figure 4.
Discharge coefficient C_d as a function of L_2/d_2 ratio.

in C_d. The discharge coefficient is weakly dependent on the length of the orifice which can be explained by the dominance singular pressure losses generated at the entrance over regular pressure drops provoked by wall friction. These results have been shown experimentally [33] and numerically [34].

For the low values of Re, we must take into account the linear and singular pressure losses.

4.1.2 Influence of the lip rounding

Figure 5 shows the distribution of the vapor volume fraction for different values of r_c and for inlet pressure p_{in} = 50 bar and L_2/d_2 = 5. **Figure 5** shows the important effect of rc on the creation of cavitation within the injector. When r_c = 0, cavitation develops over almost 1/3 the nozzle length and the cavitation zone extends to the combustion chamber. The cavitation area attached to the wall will be reduced by increasing the r_c radius.

In order to study the effect of r_c/d_2 ratio on the flow characteristics, the discharge coefficient is calculated for two values of Reynolds number Re and for L_2/d_2 = 5 (**Figure 6**).

There are two zones: the first zone for r_c/d_2 < 0.1 and the second zone for r_c/d_2 > 0.1. In the first zone, the coefficient of discharge varies linearly with inlet roundness for the two values of Re.

In the second zone, the discharge coefficient remains constant for any value of r_c/d_2. This result is consistent with the literature [35, 36].

4.2 Transient flow

In this section the transient simulations results are presented. **Figure 7** represents the evolution of vapor volume fraction at different times for L_2/d_2 = 5. The inlet corner radius r_c is equal to zero. The injection pressure p_{in} = 1000 bar and the exit pressure p_{out} = 50 bar.

The cavitation appears in the vicinity of the sharp edge for a time of the order of 0.6 μs. Then, the cavitation pocket elongates progressively through the orifice and reaches the nozzle exit at t = 3 μs. This result agrees well with the numerical

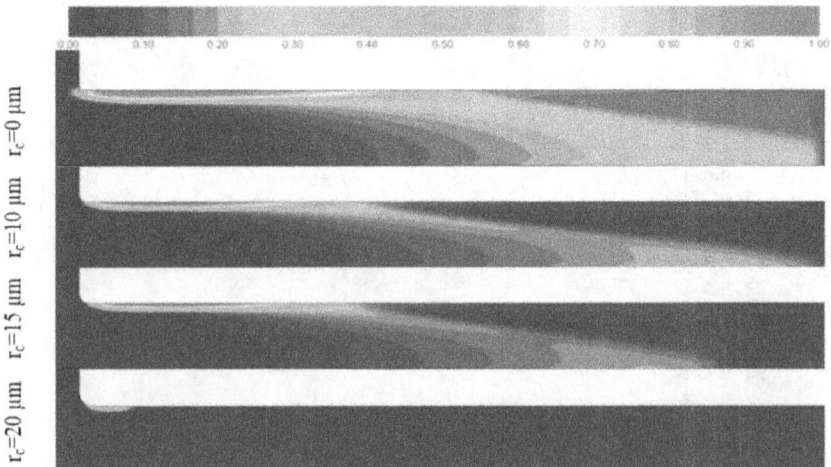

Figure 5.
Distribution of vapor fraction for several radius r_c with inlet pressure p_{in} = 50 bar and p_{out} = 1 bar.

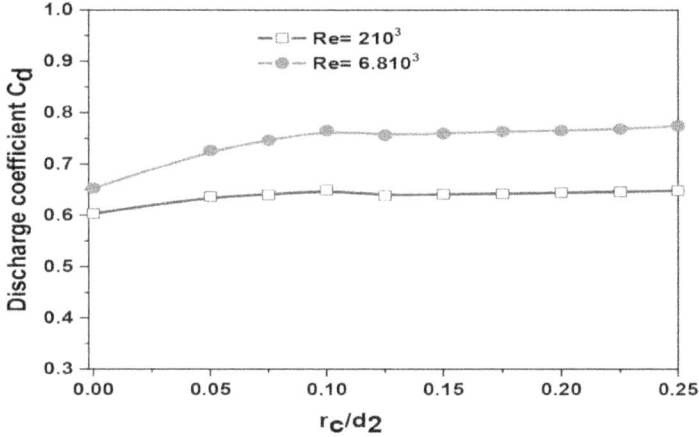

Figure 6.
Discharge coefficient as a function of rc/d2 ration for Re = 2 × 10³ and Re = 6.8 × 10³.

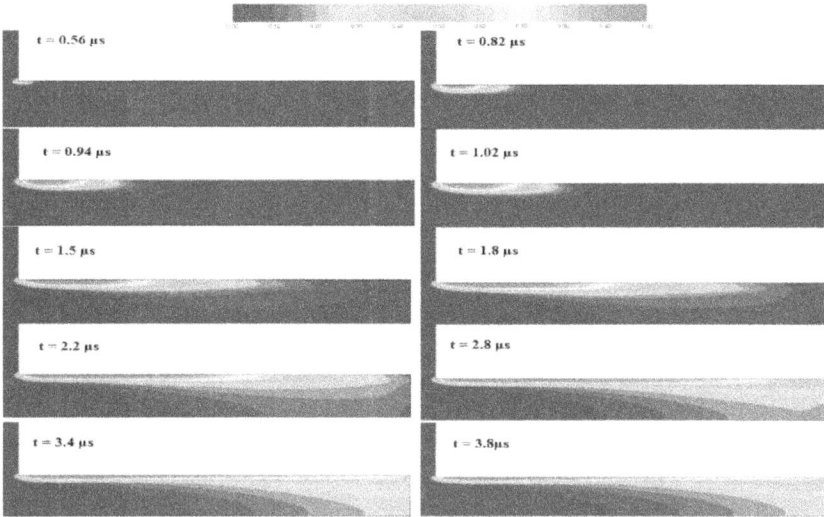

Figure 7.
Transient evolution of vapor volume fraction.

simulation obtained by Dumont et al. [37] in a similar injector and by experimental visualization and measurements Ohrn et al. [33].

Figure 8 depicts the axial profile of the mixture density near the wall at t = 1.5 μs and t = 3.8 μs. Upstream of the corner the fluid is at liquid state.

At the sharp edge, the density exhibits an important decrease due to the pressure drop and the appearance of cavitation. In this region, the mixture density contains mainly fuel vapor.

Downstream the corner, the mixture density increases and the vapor fraction decreases, owing to the collapse and the breakup of the cavitation along the orifice.

At the nozzle exit, the cavitation is present for t = 3.8 μs and therefore the mixture density is smaller than the liquid density ρ_l. Thus the spray that leaves the injector and penetrates into the combustion chamber is formed not only by fuel liquid but also by fuel vapor.

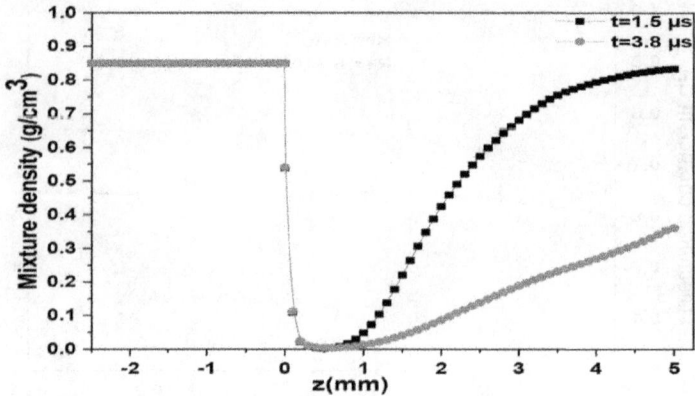

Figure 8.
Mixture density profile at t = 1.5 μs and t = 3.8 μs near the nozzle wall.

4.3 Entropy production

The entropy generation is a measure of the degree of irreversibility. It is a method for optimizing thermal and fluidic systems. It could be agreed for specific applications.

4.3.1 Case of single phase and laminar flow

4.3.1.1 Mesh sensitivity

In this part, we study the effect of the mesh on the axial component of the velocity field. The studied geometry is discretized in n × m meshes of rectangular shape. The calculation is carried out using Ansys software based on the finite volume method.

In order to assess the accuracy of the numerical method, we performed several studies based on systematic mesh refinement until there were negligible changes in the variation of the axial velocity. **Table 2** present the different mesh used in this simulation for single phase laminar flow inside the diesel injector and for two Reynold values Re = 381 and 1000, respectively.

Figure 9(a, b) shows the axial velocity profile for different mesh and for two Reynolds number Re = 381 and 1000, respectively. From this figure, the axial velocity exhibits oscillations for the first and the second mesh. This oscillation shows that the solution is not stable and that convergence has not yet been achieved. The last mesh (1453 meshes) seems the best since it shows that convergence has indeed been achieved.

Mesh	Mesh number (Re = 381)	Mesh number (Re = 1000)
Mesh 1	61	457
Mesh 2	137	1445
Mesh 3	391	5789
Mesh 4	1453	23156

Table 2.
Mesh used in this simulation.

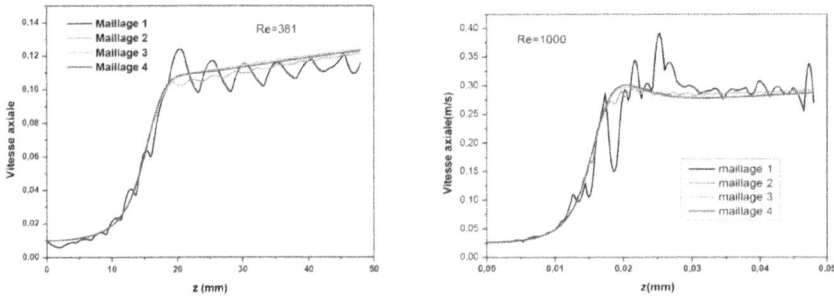

Figure 9.
Axial velocity profile for several mesh and for two values of Re = 381 (left) and Re = 1000 (right).

4.3.1.2 Entropy production

Figure 10 shows that the entropy generation increases from zero at the center of the channel to a maximum value on the wall. This trend indicates that the maximum entropy produced on the wall is mainly due to the irreversibility of fluid friction contributed by the wall velocity gradient near top while towards the center of the channel with zero velocity gradient, the generation of entropy due to friction of the fluid.

The total entropy is obtained by taking the integral of the local entropy over the total volume, given by the following expression:

$$S_{tot} = \iiint S dV \qquad (16)$$

Figure 11 illustrates the total entropy as a function of the inlet injection velocity v0. Its clear that the total entropy production varies quadratically with the inlet velocity.

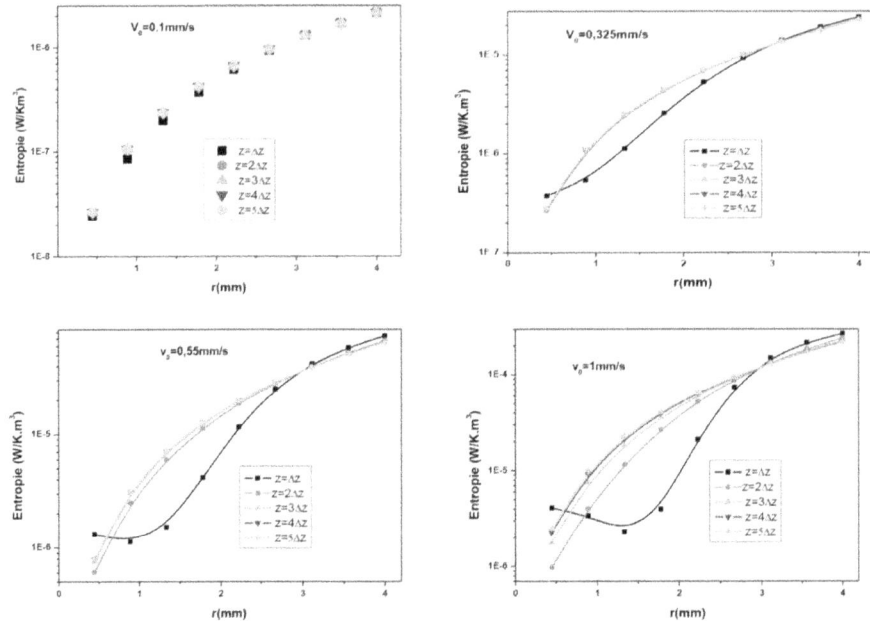

Figure 10.
Local entropy profile inside the orifice for different values of average inlet velocity v_o = 0.1, 0.325, 0.55 et 1 mm/s.

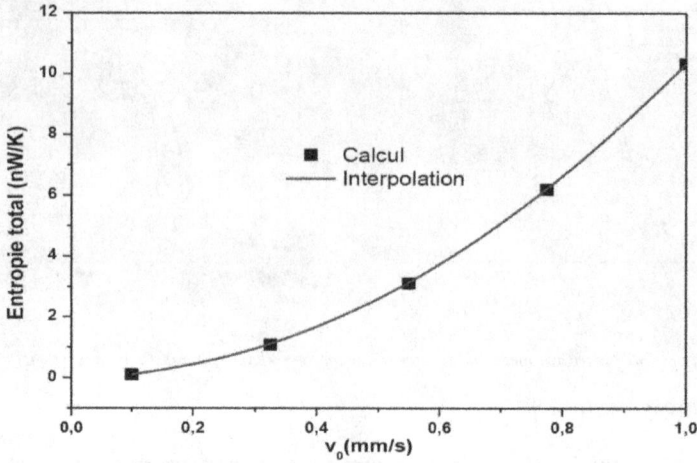

Figure 11.
Evolution of total entropy production as a function of inlet velocity.

The interpolation in Lagrange polynomial of degree 2 of S_{tot} as a function of inlet velocity v_0:

$$S_{tot}(nW/K) = C_0 + C_1 \times v_0 + C_2 \times v_0^2 \tag{17}$$

where C_0 = 0.0078 nW/K, C_1 = -0.08 nW.s/K.m and C_2 = 10.4 nW.s^2/K.m^2.

4.3.2 Case of single phase and turbulent flow

The same geometry already seen in the previous part will be used in this part. However, the flow will be considered turbulent. The main objective of this study is to simulate the entropy generation within the injector for high injection pressure taking into account the turbulent behavior of the flow. The mathematical model used consists of the Navier–Stokes equations coupled with the k-ε turbulence model taking into account the following assumptions:

- Steady flow;

- Incompressible fluid;

- Newtonian fluid;

- Isothermal flow.

Water with a density ρ = 1000 kg/m^3 and a dynamic viscosity μ = 10^{-3} Pa.s is used as a fluid.

For the velocity field and the pressure, the same boundary conditions that will be used previously at the inlet and outlet of the injector. On the other hand, at the level of the walls we use a wall law.

For the k-ε turbulence model, the boundary conditions used are as follows:

- On the axis of symmetry (r = 0)

$$\frac{\partial k}{\partial r} = 0 \ et \ \frac{\partial \varepsilon}{\partial r} = 0 \tag{18}$$

- At the inlet

$$l = C_\mu \frac{k^{3/2}}{\varepsilon} \approx 3.8\% D_h \qquad (19)$$

$$I = 0.16(Re)^{-1/8} \qquad (20)$$

- At the outlet

$$\overrightarrow{n}.\overrightarrow{gradk} = 0, \overrightarrow{n}.\overrightarrow{grad\varepsilon} = 0 \qquad (21)$$

In order to assess the accuracy of the numerical method, we performed an automatic adaptive mesh refinement. The mesh adaptation is done automatically in areas with a significant gradient (pressure, velocity, turbulent kinetic energy, kinetic energy dissipation rate).

Figures 12–15 illustrate the radial profiles of local entropy within the orifice for different axial positions. **Figures 12(a)–15(a)** show the radial entropy profiles which arise from dissipation due to the mean flow movement for various inlet velocity u_0 ranging from 10 m/s to 60 m/s. The logarithmic representation of the average viscous dissipation is shown in **Figures 12–15**. For such a single-phase viscous flow, which is both laminar and turbulent, the viscous dissipation is a function of the velocity gradient. Near the wall vicinity, the velocity gradient has a maximum value. Axially, the local entropy is important at the entrance of the orifice where the velocity gradient is important. We also notice that laminar entropy increases considerably with the inlet volocity.

Figures 12(b)–15(b) show the irreversibility due to the Reynolds shear stress resulting from the velocity fluctuation. This irreversibility can also be interpreted as

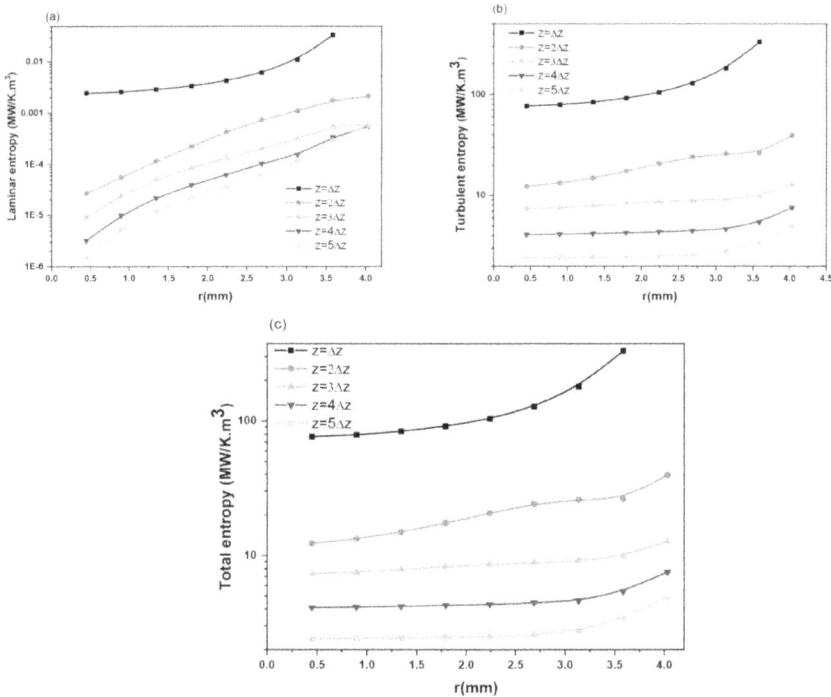

Figure 12.
Radial entropy profile at different position along the z-axis for inlet velocity $v_o = 10$ m/s. (a) Laminar entropie, (b) turbulent entropie et (c) total entropy.

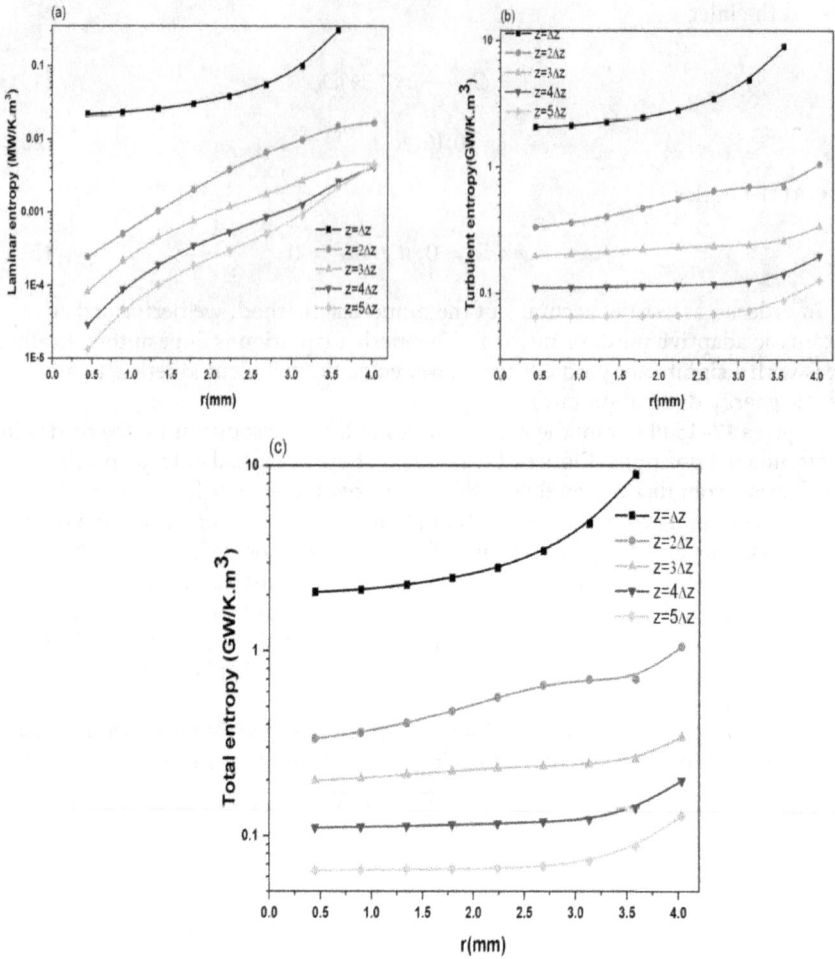

Figure 13.
Radial entropy profile at different position along the z-axis for inlet velocity v_o = 30 m/s. (a) Laminar entropie, (b) turbulent entropie et (c) total entropy.

the work done by force in the direction of the flow. From **Figures 12(a,b)-15(a,b)**, we can conclude that the entropy due to shear stress is dominant.

Figures 12(c)–15(c) illustrate the radial profiles of the total local entropy (laminar entropy and turbulent entropy) for different axial positions. As it is clear, the total entropy is important at the level of the constriction zone having a large velocity fluctuation.

Figure 16 show the local entropy distribution inside the orifice for various inlet velocity 10, 30, 50 and 60 m/s, respectively.

It is clear that the entropy is maximum in the vicinity of the wall at the contraction zone having a large velocity gradient. The greater the injection velocity, the more the irreversibility zone spreads out.

Figure 17 illustrates the total entropy as a function of the inlet injection velocity v_0. Its clear that the total entropy production varies quadratically with the inlet velocity.

The interpolation in Lagrange polynomial of degree 2 of S_{tot} as a function of inlet velocity v_0:

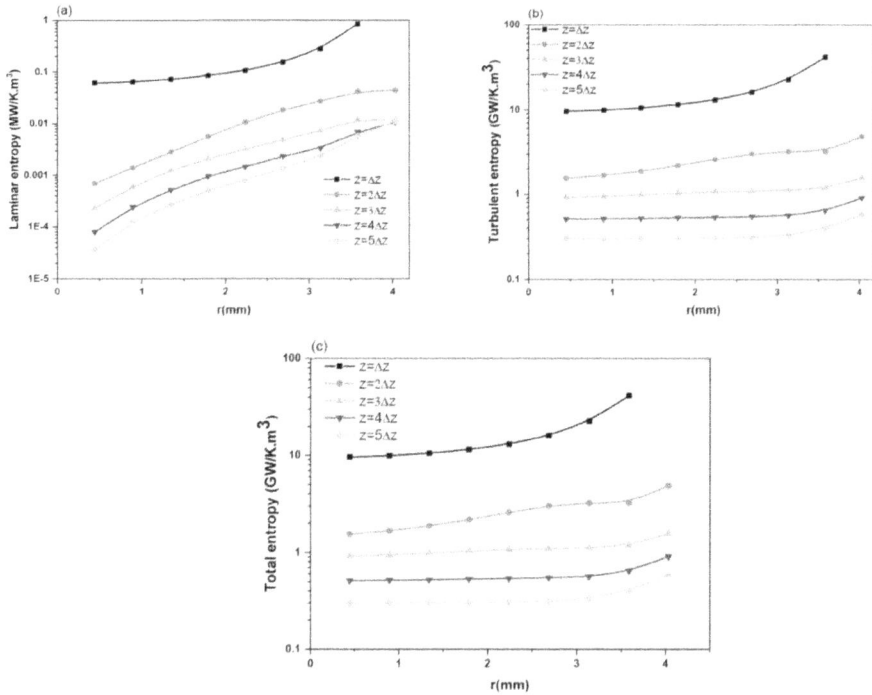

Figure 14.
Radial entropy profile at different position along the z-axis for inlet velocity v_o = 50 m/s. (a) Laminar entropie, (b) turbulent entropie et (c) total entropy.

$$S_{tot}(MW/K) = C_0 + C_1 \times v_0 + C_2 \times v_0^2 \qquad (22)$$

where C_0 = 0.29 MW/K, C_1 = -0.037 MW.s/K.m and C_2 = 0.0012 MW.s^2/K.m^2.

4.3.3 Case of two-phase turbulent cavitating flow

Our study is based on the same configuration seen in the previous sections to study the entropy generation. In fact, in this part we take into account the effect of cavitation within the orifice of the diesel injector. This will allow us to make an exhaustive study of the topology of the flow associated with the cavitation that has appeared in the injector. For this, we take into account the homogeneous mixture approach for the modeling of two-phase flows. The fluid used in this study is therefore water, the properties of which are illustrated in the following **Table 3**.

To analyze entropy generation within the injector, we have studied the effect of the injection pressure. In this case, simulations were carried out for inlet pressure varying between 1.9 bar and 1000 bar and for a fixed downstream pressure of 0.95 bar. **Figure 18** illustrates the vapor fraction and local entropy distributions inside the injector for various values of cavitation number K.

We notice from **Figure 18** (left) that the state of the fluid changes as it enters the orifice and as the injection pressure is increased. The change in the fluid state is due to the cavitation phenomenon that is created for a significant local depression. The 2D results for the vapor fraction show that cavitation is triggered when K ≈ 1.45 confirms the previous results.

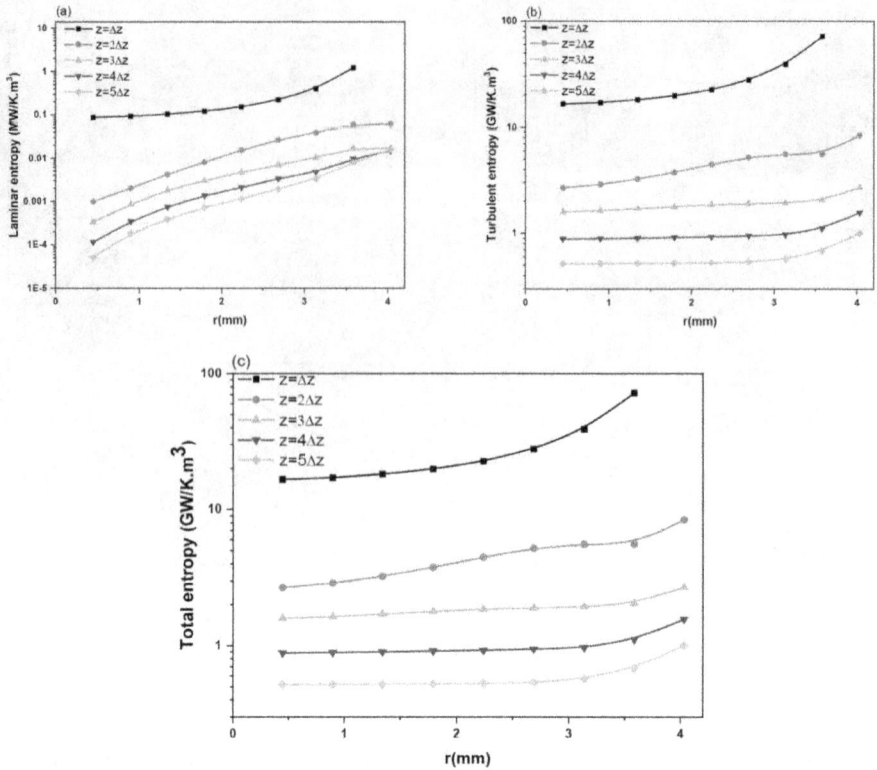

Figure 15.
Radial entropy profile at different position along the z-axis for inlet velocity v_o = 60 m/s. (a) Laminar entropie, (b) turbulent entropie et (c) total entropy.

In the center of the orifice, the fluid is formed by a dense core of fluid (liquid) and a dispersed phase (vapor), which is analogous to the experimental results of Yan and Thorpe [38]. For high injection pressures, the cavitation zone extends to the outlet leading to the formation of the hydraulic flip.

According to the results of **Figure 19** (right) show the local entropy distribution for different K values. It is clear that the entropy generation takes place near the orifice edge with a very high intensity. These results can be explained by the effect that near the wall, the radial component is important. Thus, the low pressure zone essentially gives rise to the formation of bubbles and the recirculation zone is very limited. The velocity gradient in the recirculation zone is very important causing viscous effects. This trend indicates that the maximum entropy produced near the edge of the orifice is mainly due to the irreversibility of fluid friction contributed by the velocity gradient due to the abrupt change of injector section.

For $z \geq D_1$, the flow is strongly developed. The turbulent velocity components promote the transfer of momentum between adjacent layers of the fluid and tend to reduce the average velocity gradient and subsequently a decrease in the degree of irreversibility.

Figure 19 shows the evolution of total entropy as a function of the number of cavitation K. These results prove that the degree of irreversibility is proportional to the injection pressure. For high injection pressures, entropy production is important.

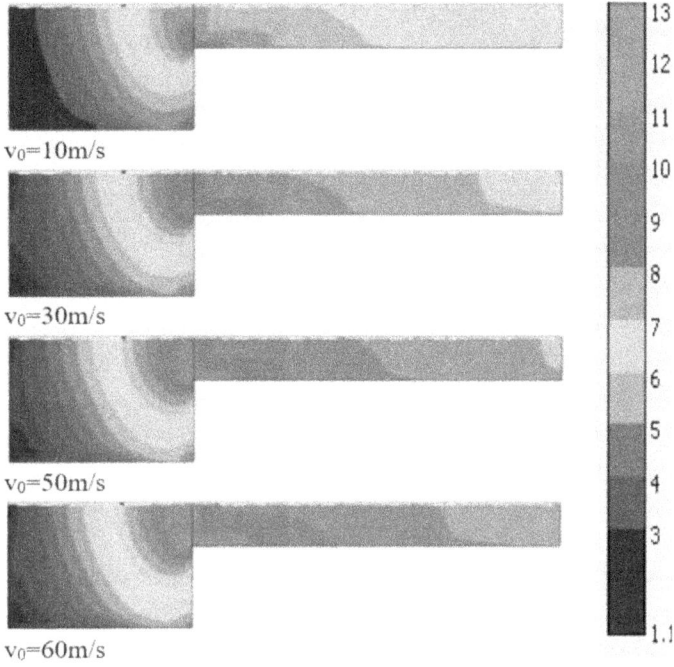

Figure 16.
Total entropy distribution for various inlet velocities v0 = 10, 30, 50 and 60 m/s. the results are presented in logarithmic decimal scale.

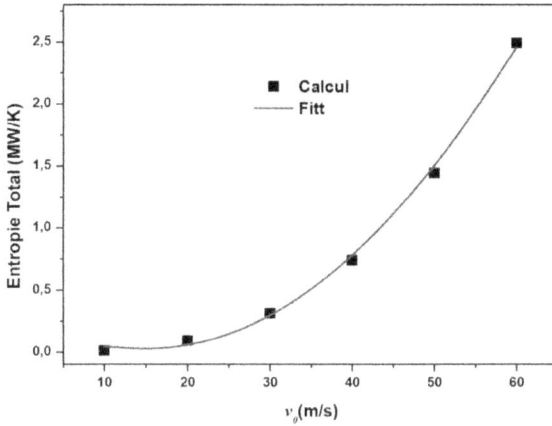

Figure 17.
Evolution of total entropy production as a function of inlet velocity.

	liquid	vapor
Density (kg/m^3)	1000	0.02558
Dynamic viscosity (kg/m-s)	0.001	1.2610^{-6}
Surface tension (kg/s^2)	0.0717	—
Vapor pressure (Pa)	3540	—

Table 3.
Water properties.

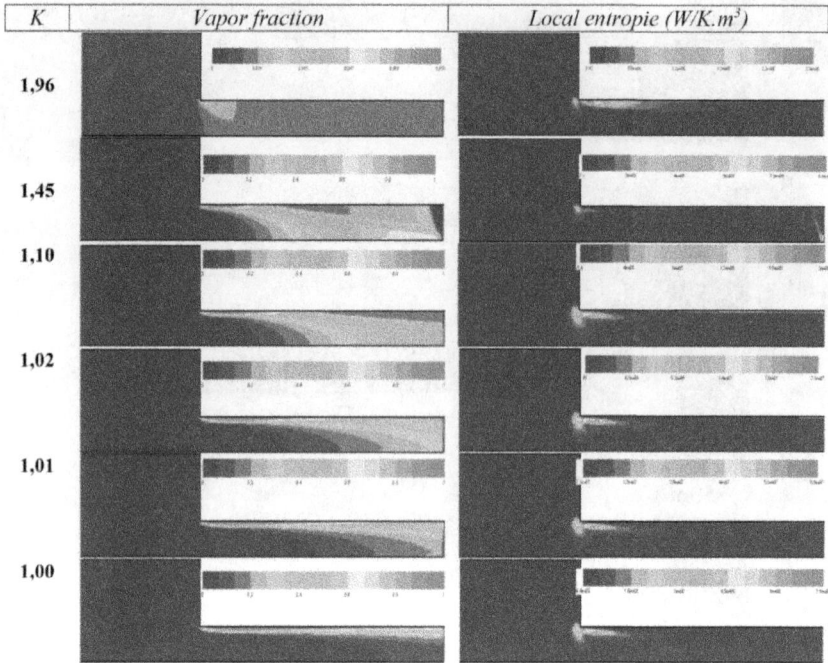

Figure 18.
Distribution of vapor fraction (left) and entropy (right) for various cavitation number values.

5. Conclusion

A numerical simulation of the phenomenon of cavitation in a 2D-axisymmetric configuration of a diesel injector nozzle has been presented. A turbulent mixture model has been adopted to simulate the multiphase flow. The study is carried out in two regimes, namely stationary and transient. The effect of the geometric parameters of the injector (e.g orifice length and the corner radius) on the cavitation phenomenon inside the nozzle has been studied.

Unsteady simulations have also been carried out. The space–time evolution of the vapor fraction is analyzed.

In the other hand, the entropy production inside the diesel injector is analyzed. Firstly, the local entropy distribution in the orifice is studied for a single phase and laminar flow under several average inlet velocity. In the second part, we analyzed the entropy production for a single phase and trubulent flow. In the same way, the effect of average inlet velocity on the entropy production is studied. Finally, the entropy production is studied for two-phase, turbulent and cavitating flow. In this case, the effet of cavitation number on local entropy production in the orifice is analyzed.

The main conclusions from the present study can be summarized as follows:

a. The study shows that the discharge coefficient is weakly influenced by the length of the orifice, especially for high injection pressures. However, the radius of the wedge has a large effect on the intensity and distribution of cavitation along the injection nozzle.

b. It appears that the spray leaving the orifice and entering into the combustion chamber contains liquid and vapor. Hence, the cavitation is expected to have an effect on the atomization and the combustion processes.

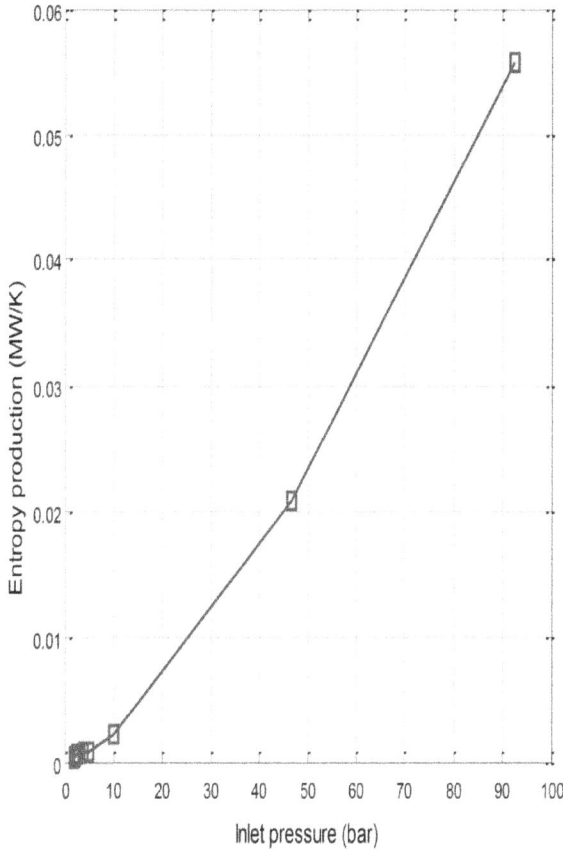

Figure 19.
Total entropy production in decimal logarithm as a function of cavitation number K.

  c. The entropy production is important near the wall and increases whith increasing the average inlet velocity.

  d. The turbulent entropy production dominates over laminar entropy production for a single phase and trubulent flow.

  e. The maximum local entropy production is localized near the wall at the entrance of the orifice.

  f. The numerical results shows the increase of total entropy generation by increasing the injection pressure.

Author details

Fraj Echouchene* and Hafedh Belmabrouk
Department of Physics, Laboratory of Electronics and Microelectronics, Faculty of
Science of Monastir, University of Monastir, Tunisia

*Address all correspondence to: frchouchene@yahoo.fr

IntechOpen

References

[1] Arcoumanis, C., M. Gavaises, H. Flora, and H. Roth. Visualisation of cavitation in diesel engine injectors. Mécanique & industries. 2001. **2**(5): 375-381.

[2] Habchi, C., N. Dumont, and O. Simonin. Multidimensional simulation of cavitating flows in diesel injectors by a homogeneous mixture modeling approach. Atomization and sprays. 2008. **18**(2).

[3] Payri, F., V. Bermúdez, R. Payri, and F. Salvador. The influence of cavitation on the internal flow and the spray characteristics in diesel injection nozzles. Fuel. 2004. **83**(4-5): 419-431.

[4] Ishimoto, J. and K. Kamijo. Numerical simulation of cavitating flow of liquid helium in venturi channel. Cryogenics. 2003. **43**(1): 9-17.

[5] Medvitz, R.B., R.F. Kunz, D.A. Boger, J.W. Lindau, A.M. Yocum, and L. L. Pauley. Performance analysis of cavitating flow in centrifugal pumps using multiphase CFD. J. Fluids Eng. 2002. **124**(2): 377-383.

[6] Desai, V., M.A. Shenoy, and P.R. Gogate. Degradation of polypropylene using ultrasound-induced acoustic cavitation. Chemical Engineering Journal. 2008. **140**(1-3): 483-487.

[7] Saletes, I., B. Gilles, and J.-C. Bera. Promoting inertial cavitation by nonlinear frequency mixing in a bifrequency focused ultrasound beam. Ultrasonics. 2011. **51**(1): 94-101.

[8] Yuan, W. and G.n.H. Schnerr. Numerical simulation of two-phase flow in injection nozzles: Interaction of cavitation and external jet formation. J. Fluids Eng. 2003. **125**(6): 963-969.

[9] Andriotis, A., M. Gavaises, and C. Arcoumanis. Vortex flow and cavitation in diesel injector nozzles. Journal of Fluid Mechanics. 2008. **610**: 195-215.

[10] Jia, M., D. Hou, J. Li, M. Xie, and H. Liu, A micro-variable circular orifice fuel injector for HCCI-conventional engine combustion-Part I numerical simulation of cavitation. 2007, SAE Technical Paper.

[11] Kim, J.-H., H.-D. Kim, K.-A. Park, S. Matsuo, and T. Setoguchi. A fundamental study of a variable critical nozzle flow. Experiments in Fluids. 2006. **40**(1): 127-134.

[12] Payri, R., F. Salvador, J. Gimeno, and J. De la Morena. Study of cavitation phenomena based on a technique for visualizing bubbles in a liquid pressurized chamber. International Journal of Heat and Fluid Flow. 2009. **30** (4): 768-777.

[13] Suh, H., S. Park, and C. Lee. Experimental investigation of nozzle cavitating flow characteristics for diesel and biodiesel fuels. International Journal of Automotive Technology. 2008. **9**(2): 217-224.

[14] Suh, H.K. and C.S. Lee. Effect of cavitation in nozzle orifice on the diesel fuel atomization characteristics. International journal of heat and fluid flow. 2008. **29**(4): 1001-1009.

[15] Zhang, X., Z. He, Q. Wang, X. Tao, Z. Zhou, X. Xia, and W. Zhang. Effect of fuel temperature on cavitation flow inside vertical multi-hole nozzles and spray characteristics with different nozzle geometries. Experimental Thermal and Fluid Science. 2018. **91**: 374-387.

[16] Cristofaro, M., W. Edelbauer, P. Koukouvinis, and M. Gavaises. A numerical study on the effect of cavitation erosion in a diesel injector. Applied Mathematical Modelling. 2020. **78**: 200-216.

[17] Echouchene, F., H. Belmabrouk, L. Le Penven, and M. Buffat. Numerical simulation of wall roughness effects in cavitating flow. International Journal of Heat and Fluid Flow. 2011. **32**(5): 1068-1075.

[18] He, Z., H. Zhou, L. Duan, M. Xu, Z. Chen, and T. Cao. Effects of nozzle geometries and needle lift on steadier string cavitation and larger spray angle in common rail diesel injector. International Journal of Engine Research. 2020: 1468087420936490.

[19] Li, D., S. Liu, Y. Wei, R. Liang, and Y. Tang. Numerical investigation on transient internal cavitating flow and spray characteristics in a single-hole diesel injector nozzle: A 3D method for cavitation-induced primary break-up. Fuel. 2018. **233**: 778-795.

[20] Santos, E.G., J. Shi, M. Gavaises, C. Soteriou, M. Winterbourn, and W. Bauer. Investigation of cavitation and air entrainment during pilot injection in real-size multi-hole diesel nozzles. Fuel. 2020. **263**: 116746.

[21] Xue, F., F. Luo, H. Cui, A. Moro, and L. Zhou. Numerical analyses of transient flow characteristics within each nozzle hole of an asymmetric diesel injector. International Journal of Heat and Mass Transfer. 2017. **104**: 18-27.

[22] Zhang, L., Z. He, W. Guan, Q. Wang, and S. Som. Simulations on the cavitating flow and corresponding risk of erosion in diesel injector nozzles with double array holes. International Journal of Heat and Mass Transfer. 2018. **124**: 900-911.

[23] Echouchene, F. and H. Belmabrouk. Computation of Cavitating Flows in a Diesel Injector. in IOP Conference Series: Materials Science and Engineering. 2010. IOP Publishing.

[24] Torelli, R., S. Som, Y. Pei, Y. Zhang, and M. Traver. Influence of fuel properties on internal nozzle flow development in a multi-hole diesel injector. Fuel. 2017. **204**: 171-184.

[25] Shi, H., M. Li, P. Nikrityuk, and Q. Liu. Experimental and numerical study of cavitation flows in venturi tubes: From CFD to an empirical model. Chemical Engineering Science. 2019. **207**: 672-687.

[26] Keller, A. The effect of flow turbulence on cavitation inception. ASME Fluids Eng. Div., FEDSM'97. 1997: 1-8.

[27] Coutier-Delgosha, O., R. Fortes-Patella, and J.-L. Reboud. Evaluation of the turbulence model influence on the numerical simulations of unsteady cavitation. J. Fluids Eng. 2003. **125**(1): 38-45.

[28] Dular, M., R. Bachert, B. Stoffel, and B. Širok. Experimental evaluation of numerical simulation of cavitating flow around hydrofoil. European Journal of Mechanics-B/Fluids. 2005. **24**(4): 522-538.

[29] Singhal, A.K., M.M. Athavale, H. Li, and Y. Jiang. Mathematical basis and validation of the full cavitation model. J. Fluids Eng. 2002. **124**(3): 617-624.

[30] Wang, C., Y. Zhang, Z. Yuan, and K. Ji. Development and application of the entropy production diagnostic model to the cavitation flow of a pump-turbine in pump mode. Renewable Energy. 2020. **154**: 774-785.

[31] Ferziger, J.H., M. Perić, and R.L. Street, Computational methods for fluid dynamics. Vol. 3. 2002: Springer.

[32] Giannadakis, E., M. Gavaises, and C. Arcoumanis. Modelling of cavitation in diesel injector nozzles. Journal of Fluid Mechanics. 2008. **616**: 153-193.

[33] Ohrn, T., D.W. Senser, and A.H. Lefebvre. Geometrical effects on

discharge coefficients for plain-orifice atomizers. Atomization and Sprays. 1991. **1**(2).

[34] Schmidt, D.P., C.J. Rutland, and M. L. Corradini. A fully compressible, two-dimensional model of small, high-speed, cavitating nozzles. Atomization and sprays. 1999. **9**(3).

[35] Martynov, S., D. Mason, and M. Heikal. Numerical simulation of cavitation flows based on their hydrodynamic similarity. International Journal of Engine Research. 2006. 7(3): 283-296.

[36] Nurick, W. Orifice cavitation and its effect on spray mixing. 1976.

[37] Dumont, N., O. Simonin, and C. Habchi. Numerical simulation of cavitating flows in diesel injectors by a homogeneous equilibrium modeling approach. http://resolver. caltech. edu/ cav2001: sessionB6. 005. 2001.

[38] Yan, Y. and R. Thorpe. Flow regime transitions due to cavitation in the flow through an orifice. International journal of multiphase flow. 1990. **16**(6): 1023-1045.

Chapter 12

Effects of Mass-Loading on Performance of the Cyclone Separators

Vikash Kumar and Kailash Jha

Abstract

The concentration of dust particles in the incoming air has a significant effect on the performance of cyclone separators. In the proposed work, four particle concentrations (0.05, 0.1, 0.2 and 0.3 kg/kg) in the dust stream have been mathematically investigated using both one-way and two-way coupling. A fine grid is solved using the Large Eddy Simulations to capture the accurate physical phenomena. For vortices smaller than the grid size the Smagorinsky Lilly model with a Smagorinsky constant, C_s of 0.1 has been applied. The collection efficiencies of the four cases are presented and the differences between the one-way and two-way coupled simulation approaches are highlighted in the present study. The two-way coupled simulations show an increase in the collection efficiency from 62% to 70% on increasing the solid loadings from 0.05 to 0.3 with marginal increase in pressure drop of 4.8%.

Keywords: stairmand high-efficiency cyclone, computational fluid dynamics, large Eddy simulations, solid loadings, Eulerian-Lagrangian

1. Introduction

Petrochemical, cement, mineral, powder processing industries, use cyclone separators extensively for separating or filtering fine particles from the primary phase. They are also used in boilers for recovery of particulate fuel, in vacuum cleaners and in biomedical applications for filtering microscopic bacteria from the air. In industries all over the world, cyclone separators are subjected to a variety of operating conditions at which they need to work efficiently. Therefore developing an understanding of the relation between performance and the operating conditions is important. The operating conditions mentioned here refer to the velocity, temperature of the incoming dust-laden air and the size distribution, the concentration of the particulate matter suspended in the air stream. The performance of the cyclone separator is represented by two parameters, firstly the pressure loss across the cyclone, (its dimensionless form being the Euler number) and secondly its collection efficiency over the range of particle diameters (dimensionless form being the Stokes number), fed into the cyclone. This is presented in the form of a curve plotted between the particle diameter (Stokes number) and its collection efficiency known as the grade efficiency curve. This curve brings into the picture a new parameter known as the cut-size diameter Stk_{50} which has been used extensively to provide a sense of the collection efficiency of a cyclone separator. The cut-size diameter is the particle diameter that corresponds to 50% collection efficiency on the grade efficiency curve. Physically it means that for the given operating conditions and the particular cyclone separator, particle diameters lesser than the cut-size will have less than 50% of the incoming solids getting collected in the dustbin. Whereas more than 50% of the larger particle diameters will be collected in the dust bin. This is observed in cyclone separators as the heavier particles have greater inertia and are more prone to separate from the swirling flow in the cyclone separator and move towards the walls. But the lesser diameter particles tend to move with the flow and are hence very difficult to separate. This behavior which determines whether a particle will be heavy enough to separate from the flow or be light enough to follow the flow depends on the drag exerted on the particle by the airflow and the Reynolds number of the flow. As the Reynolds number of the flow is same for all the particle sizes, the drag emerges as an important factor and is dependent on the diameter and roughness of the particle as suggested earlier that the grade efficiency curve is a function of the particle size.

Research suggests that apart from particle size, the grade efficiency curve or simply the collection efficiency for different particle sizes also shows a variation with the particle concentration inside the cyclone separator. Hoffmann et al. [1] experimented three cyclone models with different vortex finder diameters at three different inlet velocities and presented the variation of the two performance parameters with respect to solids loadings (0–0.04 kg/m^3). The relation between separation efficiency and vortex finder diameter was found to be inversely proportional as reported by Brar et al. [2]. The separation efficiency improved with inlet velocity but only up to a certain solid loading, after which the reverse was seen. They concluded that an increase in solid loadings reduced the pressure drop slightly and increased the overall collection efficiency. An interesting point observed was that at lower inlet velocity (10 m/s) the improvement in efficiency due to an increase in solid loadings was higher than that at higher inlet velocity (20 m/s). The curves crossed at a solid loading of 0.02 kg/m^3. A similar observation of obtaining a higher collection efficiency at 18 m/s than at 27 m/s was reported by Fassani and Goldstein [3] who conducted experiments at solid loadings as high as 20 kg of solids/kg of gas. They further reported that dust-laden air showed lower pressure drop values than that of dust-free air, collection efficiency improved with solid loadings up to 12 kg of solids/kg of gas, after which reduction in efficiency was seen.

It has been reported that two-way coupling simulations are necessary beyond a certain solid loading threshold to capture the reductions in the swirl [4]. It has also been reported that if the cyclone separator load exceeded the critical load [1], improvements in separation performance would be seen. The present paper performance analysis of the Stairmand High-Efficiency Cyclone (SHEC) at four different solid loadings with the help of one way and two-way coupled Large-eddy Simulations. Both one way and two-way coupled simulations are performed to enquire if they can capture the reduction in swirl and the improvements in separation performance at solid loadings of 0.05, 0.1, 0.2 and 0.3 (kg of dust per kg of air). Many studies on the effect of mass loadings on the efficiency of cyclone separators are available in literature such as [1, 4–6]. But the upper limit of the range studied in most of them is only 0.1 kg of dust per kg of air, whereas Derksen et al. [7] investigated the effects on the flow up to the range of 0.2. Although the solids loading of 0.2 has been investigated by Derksen et al. [7] the authors stated that as the simulations were computationally expensive, they could not perform extended simulations to get quantitative results on various quantities. The present paper extends the upper limit of solid loadings to 0.3 and performs a comparison of one way and two-way coupled simulation strategies. The validation of the clean gas flow pattern with experimental data [8] has been done, this helps emphasize the particle modeling approach and its accuracy.

2. Numerical simulations

The present gas-particle simulations are performed using the Eulerian-Lagrangian approach, the gas flow field is determined using a Eulerian approach and the particles' trajectories are calculated using a Lagrangian approach. The fluid medium is considered as a continuum for which the Navier-Stokes equation pertaining to the coordinate directions is solved and the particle or dispersed phase trajectories are tracked in the developed gas flow field. The particle equation of motion is used to calculate the particle velocity from the instantaneous gas velocity.

2.1 Gas flow

The gas flow inside the cyclone separator is modeled using Large Eddy Simulations (LES), which has been successfully applied to flows involving cyclone separators [9–15]. LES has the potential to resolve most of the energy-carrying larger eddies directly, whereas smaller energy eddies are modeled using a sub-grid scale (SGS) model. The sub-grid scale refers to scales smaller than the grid size. As the grid size determines the scales of eddies that will be directly resolved, a smaller grid size results in higher accuracy at the cost of higher computational time. The Smagorinsky-Lilly model [16] with a Smagorinsky constant of 0.1 has been used for modeling the sub-grid scale stresses. The sub-grid scale stresses are calculated employing the Boussinesq hypothesis [17] similar to that in the Reynolds averaged Navier-Stokes (RANS), model.

As turbulence introduces erratic velocity fluctuations in the flow field, appropriate models are needed to capture the turbulent characteristics of the physical flow field into the numerical calculations. Many models have been developed over time to approximate the turbulent behavior of fluids, Spalart-Allmaras model, k-ε model and k-ω models are some that use the Boussinesq hypothesis. These models are relatively less expensive in a computational sense but possess shortcomings in many complex flows. Reynolds Stress equation model is another approach of modeling turbulence which uses three additional equations for determining the Reynolds stresses is considered in the present study. Accuracy of the Reynolds Stress model in the context of flows in cyclone separators has been well established in the literature. The Reynolds averaged continuity and Navier-Stokes equations are given as [2]:

$$\frac{\partial \overline{u_i}}{\partial x_i} = 0 \tag{1}$$

$$\frac{\partial \overline{u_i}}{\partial t} + \overline{u_i}\frac{\partial \overline{u_j}}{\partial x_j} = -\frac{1}{\rho}\frac{\partial \overline{P}}{\partial x_i} + \nu\frac{\partial^2 u_i}{\partial^2 x_j} + \frac{\partial}{\partial x_j}\tau_{ij} \tag{2}$$

where $\overline{u}$ is the mean velocity, x is the coordinate direction, $\overline{P}$ the average static pressure, ρ the gas density, ν the kinematic viscosity and τ_{ij} the Reynolds stress tensor ($\tau_{ij} = -u_i{'}u_j{'}$) with u' as the fluctuating component of velocity, this term represents the transfer of momentum due to turbulence. Eq. (1) is the differential form for the conservation of mass for a small fluid element whereas Eq. (2) is the differential form for the conservation of momentum. The terms on the left hand side of Eq. (2) collectively describe the acceleration of a fluid element, the first term denoting the temporal acceleration and the second denoting the convective acceleration. The terms on the right hand side represent the pressure force, viscous force and the Reynolds stress tensor.

2.2 Particle motion

The particle motion inside the cyclone separator is described using a Lagrangian approach, the equation for motion of the particle is given by Clift et al. [18]:

$$\frac{\partial u_{pi}}{\partial t} = F_D\left(u_i - u_{pi}\right) + \frac{\left(\rho_p - \rho\right)g_i}{\rho_p} + F_i \tag{3}$$

Where u_{pi} is the particle velocity, first term on the right-hand side, denotes the drag force encountered by the particles due to fluid flow, the second term denotes forces due to gravitational accelerations and the third term includes all the additional forces that can be accounted for. The additional force in Eq. (3) comprises of virtual mass force, the pressure gradient force, the Brownian force and the Saffman's lift force some of which can be safely neglected for the case of particulate flow in cyclone separators [19]. The particle trajectories in the turbulent flow are solved by using the gas velocity derived from the gas flow field computed by the Large Eddy Simulations. The particles considered are small and interpenetrating in nature, which means the collisions between the particles are neglected in the present simulations. In cases where the particle concentration inside the cyclone separator is small, the exchange of momentum from the dispersed phase to the fluid phase can be neglected, this is known as a one way coupled simulation. Whereas when the exchange of momentum from the dispersed phase to the fluid phase becomes significant and is accounted for in the simulations, it is known as a two-way coupled simulation. In the present study, both one way and two-way coupled simulations have been employed. The effects of turbulence on the particles are incorporated using a discrete random walk model [20], this model simulates the interaction of the particles with discrete stylized fluid phase turbulent eddies. The turbulent eddies are characterized by Gaussian distributed random velocity fluctuations and the eddy time scale. Assuming anisotropy of the stresses, the instantaneous turbulent velocity fluctuations are given as [20]:

$$u_i' = \xi\sqrt{\overline{u_i'^2}} \tag{4}$$

Each interaction of the particle with the turbulent eddy, is considered over a time scale which is shorter of the eddy time scale and the eddy crossing time, given by:

$$\tau_e = -T_L \ln(r) \tag{5}$$

Where r is a random number between 0 and 1 and T_L is given by:

$$T_L \approx 0.15\frac{k}{\varepsilon} \tag{6}$$

2.3 Cyclone geometry

The cyclone geometry considered in the present study is the Stairmand High-Efficiency Cyclone [21], same as that used by Hoekstra [8] in experiments and by Derksen et al. [4] in large-eddy simulations using the lattice-Boltzmann discretization scheme. The choice of cyclone geometry is made keeping in mind the experimental data available facilitating validation of the simulation procedure. The cyclone geometry used for the simulations is presented in **Figure 1**. A similar obstruction with diameter 0.36 D_{exit} has been provided slightly upstream of the outflow boundary in the present geometry in order to prevent subcritical flow [11].

2.4 Discretization of the solution domain

The cyclone geometry has been meshed using ICEM CFD, a meshing tool which comes along the Ansys commercial software package which uses the multi-block meshing approach. The domain is meshed using high-quality hexahedral elements

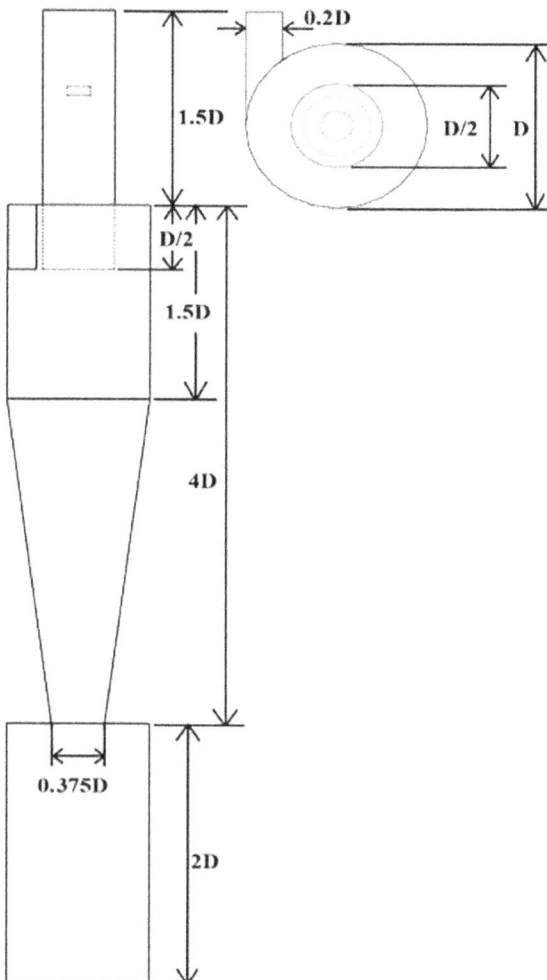

Figure 1.
Cyclone dimensions D = 0.29 m.

with the lowest element orthogonality being 0.06 and highest element aspect ratio being 31.12. For the case of LES, a small value of Δx should be used in order to minimize the small scale energy-carrying eddies. However, this can lead to a high computational cost therefore a compromise has to be made, keeping in mind the steep velocity gradients in the boundary layer. The value of Δx is taken as D/90 on the innermost block taken from [11] resulting in approximately 2×10^6 elements.

2.5 Solver settings

The present simulations have been carried on the commercial software Ansys Fluent v15.0. A pressure-based transient simulation has been conducted to capture the fluid flow inside the cyclone separator. For capturing the effects of turbulence, the Reynolds stress model with a linear pressure-strain relationship is employed, standard wall functions based on the works of Launder and Spalding [22] are used for treating the boundary layer flow. For the gas flow, air with a density of 1.225 kg/m^3 and dynamic viscosity of 1.7894×10^{-5} kg/m-s and inlet velocity of 16.1 m/s which corresponds to a Reynolds number of 28×10^5 has been taken. The inlet velocity is the same as that taken by Hoekstra [8] in experiments and by Derksen et al. [4] in simulations. The boundary conditions for the fluid phase has been specified as velocity-inlet at the inlet, outflow at the outlet and all other surfaces has been defined as walls. For the dispersed phase, the inlet boundary is prescribed as an escape for the particles, the bottom of the dustbin is prescribed as a trap and the remaining surfaces are considered as walls. The walls are set to reflect the particles with normal and tangential wall reflection coefficients as 1, facilitating elastic collisions between the walls and particles.

The discretization schemes used for pressure velocity coupling is SIMPLEC (Semi-Implicit Method for Pressure Linked Equations-Consistent), the pressure

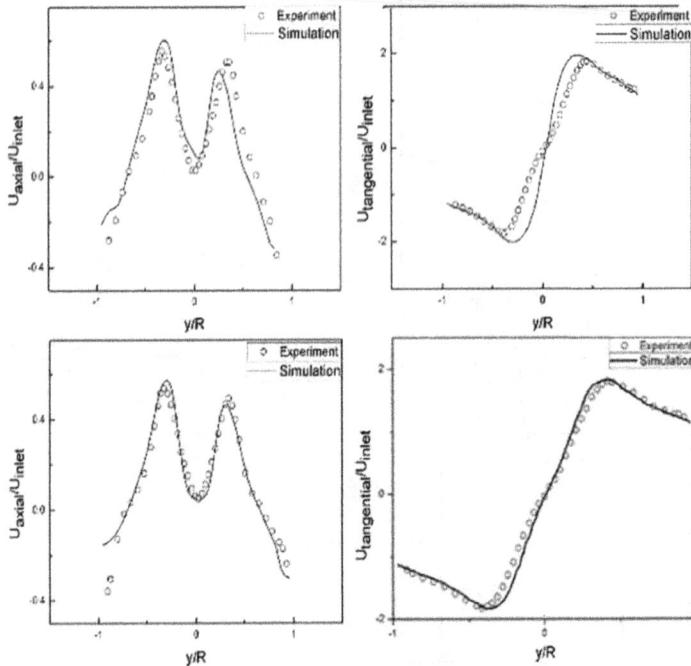

Figure 2.
Comparison of numerically simulated velocity profiles with LDA measurements of Hoekstra [8].

term in the momentum equation is discretized using PRESTO and the momentum terms using QUICK. The turbulent kinetic energy and turbulent dissipation terms are discretized using the first-order upwind scheme.

2.6 Validation

The simulation methodology used in the present study has been validated using the experimental results of Hoekstra [8]. **Figure 2** shows a comparison of the velocity profiles at two axial locations, 0.75D and 2D, generated numerically and through LDA measurements. Good agreement between the velocity profiles is seen and it can be said that the simulations are able to approximate the physical phenomenon with reasonable accuracy.

3. Results and discussion

The results of the two simulation approaches at the four solid concentrations have been analyzed in this section. The mean axial and tangential velocity, profiles at two axial locations (0.75D and 2D from the roof of the cyclone) have already been presented and validated in Section 2.3, and the same flow pattern has been used for injection of particles with different mass loadings.

3.1 Pressure drop

The pressure drop for the clean gas simulation that was run before injecting the particles amounted to 1398.01 Pa but this value increased on injection of particles. Whereas, Fassani and Goldstein [3] and Hoffmann et al. [6] reported the pressure drop to decrease on the introduction of dust particles. Moreover, this reduction was more pronounced at higher inlet velocities. The increase in pressure drop can be attributed to an increase in surface roughness of the cyclone separator walls as particles tend to accumulate towards the walls. **Table 1** shows that the increase in pressure drop for one way coupled simulations is almost same for all the mass loadings with only slight deviations. Whereas for two-way coupling, the pressure drop values are seen to rise marginally with mass loadings. This suggests that due to the increase in mass loadings, the force imparted on the fluid by the particles has a considerable effect on the pressure field inside the cyclone separator. Similar observations were made by Huang et al. [5] and the increased pressure drop was attributed to decrease in gas pressure due to increased wall friction. However, contradictory findings with respect to pressure drop have been reported in the literature [1, 3, 6], which is consistent with the present work for mass loading up to 0.1 kg/kg of gas in

Solids concentration	One-way coupling		Two-way coupling	
	Pressure drop (Pa)	Efficiency (%)	Pressure drop (Pa)	Efficiency (%)
0.05	1492.30	61.88	1486.76	62.50
0.1	1487.20	61.76	1489.24	61.86
0.2	1495.81	61.72	1514.60	66.57
0.3	1503.96	61.63	1558.33	70.31

Table 1.
Pressure drop and separation efficiency values at t = 7.45 s.

one-way coupled simulation. Particle-particle interaction may be the prime reason for reduction in pressure drop for the higher mass loading. In the proposed work mass loading is limited to 0.3 kg/kg of gas for which particle-particle interactions have not been considered. The pressure drop obtained in the proposed work for the considered mass loading ranges matches with Baskakov's model [23].

3.2 Particle motion

Four mass loadings were simulated for the SHEC, The particles were injected into the cyclone after 2.87 s of the start of the fluid flow as it is quite enough for the flow to become steady. Nine particle sizes (8.05×10^{-7}, 1.04×10^{-6}, 1.34×10^{-6}, 1.74×10^{-6}, 2.23×10^{-6}, 2.9×10^{-6}, 3.75×10^{-6}, 4.88×10^{-6}, 6.24×10^{-6} m) were injected into the cyclone and the simulations were run till $t = 6.95$ s. After which particles still remained in the separation space of the cyclone separator, but were few in number and mainly smaller in diameter. **Figure 3** shows the radial profiles of tangential velocity at three axial locations for mass loadings 0.1 and 0.3. The figure shows that for the mass loadings of 0.1, the tangential velocity distribution for one way and two-way coupled simulations is somewhat similar. This signifies that for such low mass loadings there is hardly any difference between the two simulation strategies thereby confirming the similar pressure drop and collection efficiency values obtained from the two simulation methodologies. But for mass loading of 0.3, the peak tangential velocities for the two-way coupled simulation is seen to decrease this decrease in tangential velocity indicates a change in the flow field and also a decrease in the swirl. It is also observed that this decrease in tangential velocity is more prominent in the free vortex

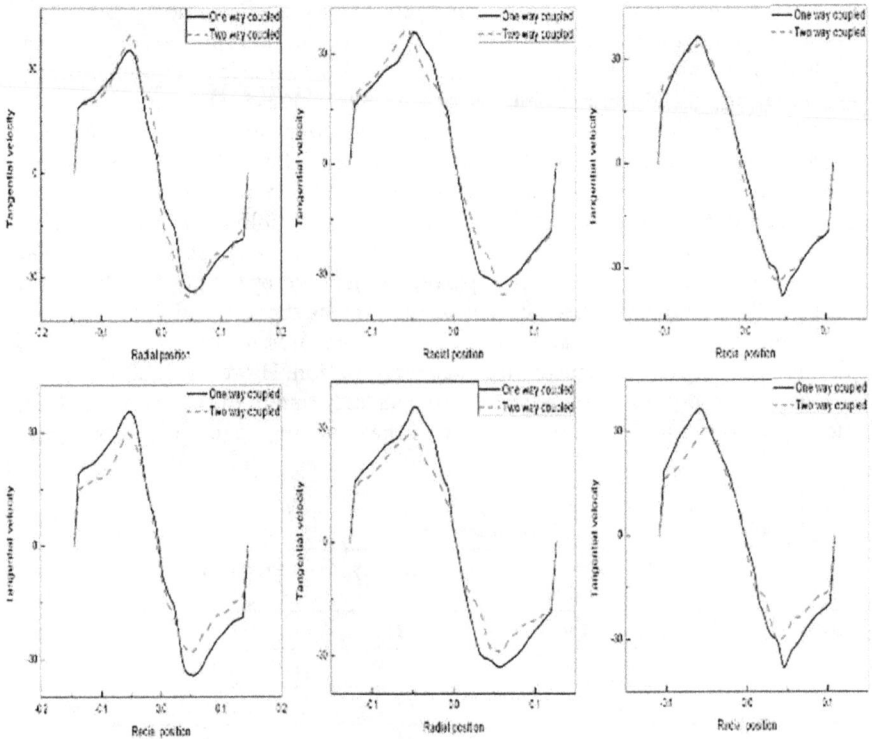

Figure 3.
Radial profiles of tangential velocity top row: mass loading 0.1, bottom row: mass loading 0.3; from left to right: axial location 0.75D, 2D, 2.5D from the cyclone roof.

region where the particles tend to accumulate. Such a behavior has also been observed by Fassani and Goldstein [3]. An interesting observation is that although the swirl reduces, the pressure drop increases though slightly for two-way coupling. This increase in pressure drop can be explained as a result of dominance of increase in wall friction due to presence of a high concentration of particles near the wall as compared to reduction in pressure drop which occurs due to reduction in tangential velocity.

The particles inside the cyclone separator for one way coupling at five instances of time for mass loadings 0.1, and 0.3 are shown in **Figures 4** and **5**. It can be clearly seen that at a later stage in the separation process, only smaller diameter particles are left out in the cyclone and the number of particles left in the cyclone after this time decreases on increasing the mass loadings. Three particle diameters at t = 4.5 s for all the mass loadings are shown in **Figures 6** and **7**. The figures confirm that the fate of larger particles is decided very early in the separation process and also that increasing the mass loadings decrease the number of smaller particles left out in the cyclone separator at a later stage. An interesting point seen in **Figures 6** and **7** is that some larger diameter particles are also found stuck in recirculation zones along the annular part of the roof of the cyclone. This gives an idea of the strength of the recirculation zones operating inside the cyclone separator. Comparison of **Figures 6** and **7** shows that at the same time, the number of particles left in the cyclone separator is less for two-way coupled simulations this suggests that the change in flow field brought about by the two-way coupled simulations not only increases the collection efficiency but also decides the fate of particles sooner than that of one way coupling. This was not observed at lower mass loadings than 0.1 but is significantly observed at only mass loadings of 0.3. Two way coupled simulations [5] also showed that the influence of particles on the gas flow field was relatively small at lower mass loadings. This gives an idea of how closely two-way coupling can approximate the real phenomena occurring inside the cyclone separator as compared to one way coupling.

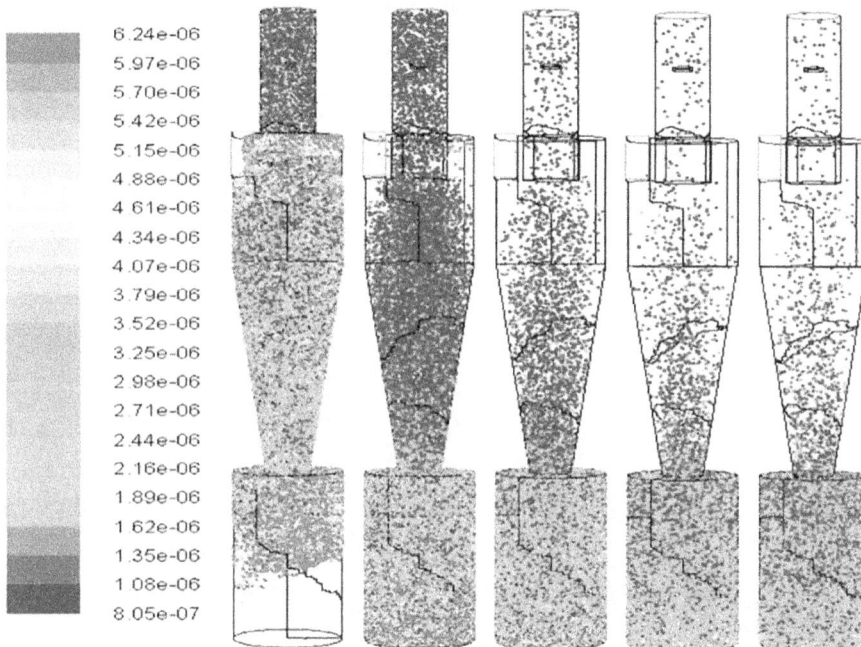

Figure 4.
Particles inside the cyclone from left to right at t = 3.2 s, 3.8 s, 4.5 s, 5.6 s, 6.9 s at mass loadings 0.1.

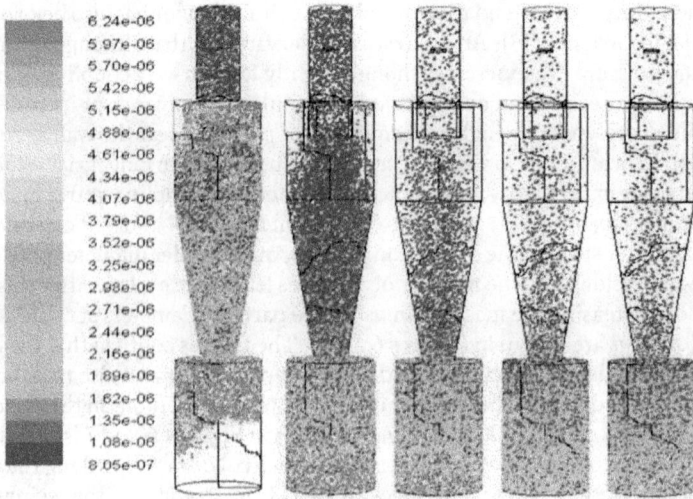

Figure 5.
Particles inside the cyclone from left to right at t = 3.2 s, 3.8 s, 4.5 s, 5.6 s, 6.9 s at mass loadings 0.3.

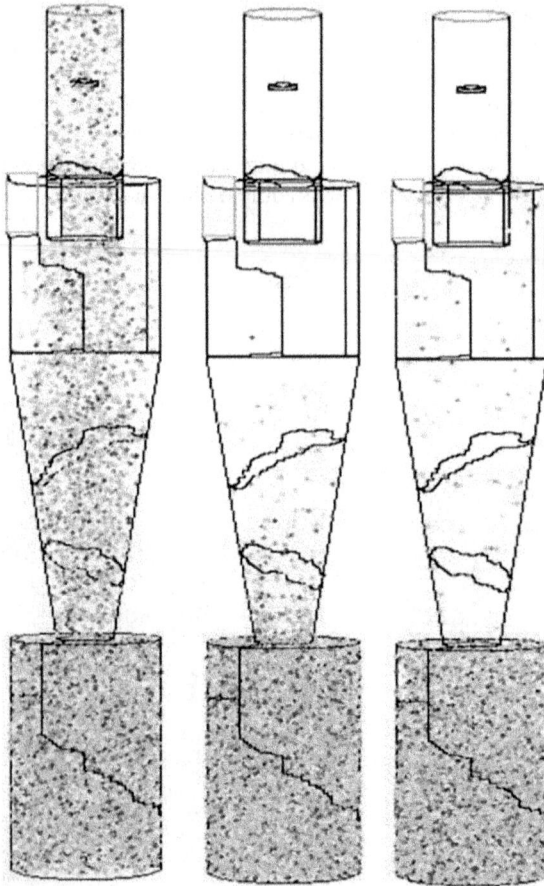

Figure 6.
Single realization of particle sizes 8.05 × 10^{-7}, 2.23 × 10^{-6}, 6.24 × 10^{-6} at t = 4.5 s, for one way coupling at mass loadings 0.3.

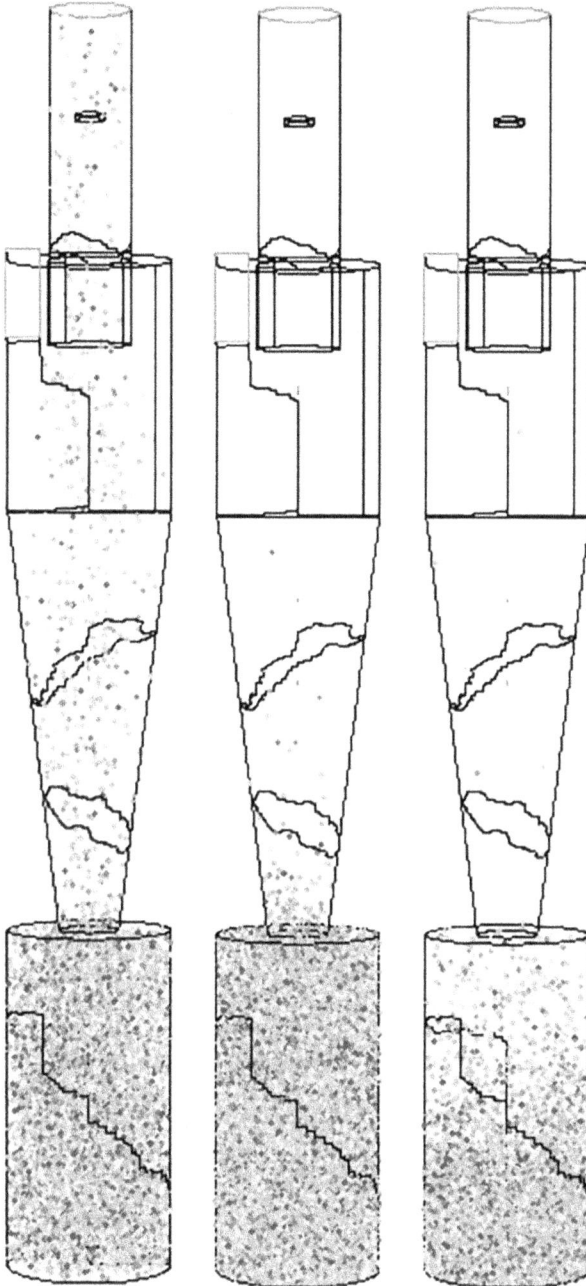

Figure 7.
Single realization of particle sizes 8.05×10^{-7}, 2.23×10^{-6}, 6.24×10^{-6} at t = 4.5 s, for two-way coupling at mass loadings 0.3.

3.3 Collection efficiency

The collection efficiency with respect to mass loadings and the simulation methodology applied at the four mass loadings are given in **Table 1**. **Figure 8** presents both the pressure drop and collection efficiency of the cyclone model at

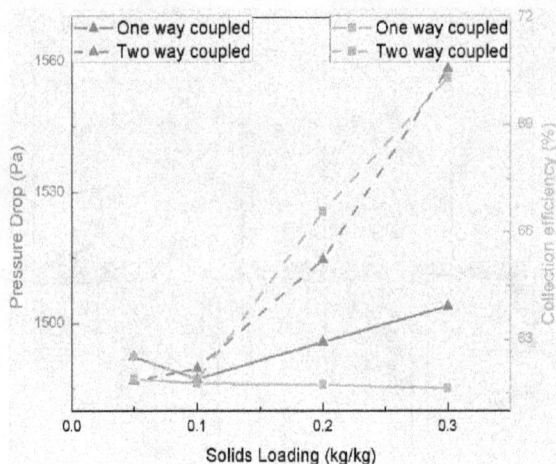

Figure 8.
Pressure drop and collection efficiency obtained at the four solid loadings.

the four solid loadings tested, it is seen that for one way coupled simulations, the pressure drop increases slightly whereas the collection efficiency is seen to decrease marginally. On the other hand, the two way coupled simulations show a marginal increase in the pressure drop and a considerable increase in collection efficiency after the solid loadings increases beyond 0.1. It is observed that one way coupled simulations predicts similar separation capability of cyclone separators at different mass loadings. A similar collection efficiency of the three cases explains the incapability of one way coupled simulations to predict the improvements pointed out by [1]. Therefore the efficiency obtained on considering one way coupled simulations can lead to incorrect estimation of the separation capability at higher mass loadings. Whereas the separation efficiency for two-way coupling can be seen to rise with mass loadings suggesting that the simulation strategy is able to account for the sweeping effect caused by larger particles. Moreover, it is observed that the two strategies of simulation provides similar results till mass loadings of 0.1 but beyond this point, differences in the pressure drop and separation efficiency for both the simulation strategies can be observed. The efficiency obtained by conducting two-way coupled simulations at higher mass loadings is more reliable than one way coupled as the latter does not account for the forces imparted on the continuous phase by the dispersed phase. The fact that any increase in mass of the particles does not have an effect back on the flow and therefore is unable to reduce the swirl.

4. Conclusions

One way and two-way coupled simulations of Stairmand High Efficiency Cyclone at four mass loadings have been performed and the results presented. It can be concluded that

- The two-way coupled simulations are seen to predict the physical phenomena better than the one-way coupled simulations after a solid loading of 0.1.

- At all the solids loadings, a similar separation capability is observed for one way coupling, but for two-way coupling, the separation efficiency and pressure drop are seen to increase with mass loadings. This fact signifies that after a certain loading condition, one way coupling is not able to correctly capture the effects of the forces that are acting on the continuous phase.

- The collection efficiency of the cyclone model is seen to increase from 62% to 70% on increasing the mass loadings from 0.05 to 0.3 with a marginal increase in pressure drop of only 4.8%.

Acknowledgements

The author would like to thank Dr. Lakhbir Singh Brar, Assistant Professor, B.I.T. Mesra, for his valuable time and help he put in this study.

Conflict of interest

The authors declare no conflict of interest.

Author details

Vikash Kumar* and Kailash Jha
Department of Mechanical Engineering, Indian Institute of Technology (ISM) Dhanbad, Dhanbad, India

*Address all correspondence to: harshvikash@gmail.com

IntechOpen

References

[1] Hoffmann AC, Arends H, Sie H. An experimental investigation elucidating the nature of the effect of solids loading on cyclone performance. Filtration & separation. 1991;**28**(3):188-193

[2] Brar LS, Sharma RP, Dwivedi R. Effect of vortex finder diameter on flow field and collection efficiency of cyclone separators. Particulate Science and Technology. 2015;**33**(1):34-40

[3] Fassani FL, Goldstein L Jr. A study of the effect of high inlet solids loading on a cyclone separator pressure drop and collection efficiency. Powder Technology. 2000;**107**(1-2):60-65

[4] Derksen JJ, Sundaresan S, Van Den Akker HEA. Simulation of mass-loading effects in gas–solid cyclone separators. Powder technology. 2006;**163**(1-2):59-68

[5] Huang AN, Ito K, Fukasawa T, Fukui K, Kuo HP. Effects of particle mass loading on the hydrodynamics and separation efficiency of a cyclone separator. Journal of the Taiwan Institute of Chemical Engineers. 2018;**90**:61-67

[6] Hoffmann AC, Van Santen A, Allen RWK, Clift R. Effects of geometry and solid loading on the performance of gas cyclones. Powder Technology. 1992;**70**(1):83-91

[7] Derksen JJ, Van den Akker HEA, Sundaresan S. Two-way coupled large-eddy simulations of the gas-solid flow in cyclone separators. AIChE Journal. 2008;**54**(4):872-885

[8] Hoekstra AJ. Gas flow field and collection efficiency of cyclone separators. TU Delft. Ph. D. Thesis, Delft University of Technology. 2000

[9] de Souza FJ, de Vasconcelos Salvo R, de Moro Martins DA. Large Eddy Simulation of the gas–particle flow in cyclone separators. Separation and Purification Technology. 2012;**94**:61-70

[10] de Souza FJ, de Vasconcelos Salvo R, de Moro Martins DA. Simulation of the performance of small cyclone separators through the use of Post Cyclones (PoC) and annular overflow ducts. Separation and Purification Technology. 2015;**142**:71-82

[11] Derksen JJ. Separation performance predictions of a Stairmand high-efficiency cyclone. AIChE Journal. 2003;**49**(6):1359-1371

[12] Elsayed K, Lacor C. The effect of cyclone vortex finder dimensions on the flow pattern and performance using LES. Computers & Fluids. 2013;**71**:224-239

[13] Shukla SK, Shukla P, Ghosh P. The effect of modeling of velocity fluctuations on prediction of collection efficiency of cyclone separators. Applied Mathematical Modelling. 2013;**37**(8):5774-5789

[14] Bogodage SG, Leung AY. CFD simulation of cyclone separators to reduce air pollution. Powder Technology. 2015;**286**:488-506

[15] Brar LS, Elsayed K. Analysis and optimization of multi-inlet gas cyclones using large eddy simulation and artificial neural network. Powder Technology. 2017;**311**:465-483

[16] Smagorinsky J. General circulation experiments with the primitive equations: I. The basic experiment. Monthly Weather Review. 1963;**91**(3):99-164

[17] Hinze JO. Turbulence. New York: McGraw-Hill Publishing Co.; 1975

[18] Clift R, Grace JR, Weber ME. Bubbles, Drops, and Particles. New York: Academic Press; 1978

[19] Hoffmann AC, Stein LE, Hoffmann AC, Stein LE. Gas Cyclones and Swirl Tubes. Vol. 56. Berlin Heidelberg New York: Springer-Verlag; 2002